A New History of Identity

Also by this author

Armstrong, D. *An Outline of Sociology as Applied to Medicine*, 4 editions

Armstong, D. *Political Anatomy of the Body: Medical Knowledge in Britain in the Twentieth Century*

Armstrong, D., Calnana, M. and Grace, J. *Research Methods in General Practice*, 3 editions

A New History of Identity

A Sociology of Medical Knowledge

David Armstrong

palgrave

First published 2002 by
PALGRAVE
Houndmills, Basingstoke, Hampshire RG21 6XS and
175 Fifth Avenue, New York, N.Y. 10010
Companies and representatives throughout the world

PALGRAVE is the new global academic imprint of St. Martin's Press LLC Scholarly and Reference Division and Palgrave Publishers Ltd (formerly Macmillan Press Ltd).

ISBN 0–333–96892–1

This book is printed on paper suitable for recycling and made from fully managed and sustained forest sources.

A catalogue record for this book is available from the British Library.

Library of Congress Cataloging-in-Publication Data

Armstrong, David, 1947 June 3–
A new history of identity: a sociology of medical knowledge / David Armstrong.
p. cm.
Includes bibliographical references and index.
ISBN 0–333–96892–1 (cloth)
1. Social medicine. 2. Identity. 3. Medicine – History. I. Title.

R133.A75 2002
306.4'61–dc21

2001058213

10 9 8 7 6 5 4 3 2 1
11 10 09 08 07 06 05 04 03 02

Printed and bound in Great Britain by
Antony Rowe, Chippenham, Wiltshire

Contents

Preface

This book has a number of aims. First, it is an attempt to write a sociology of medical knowledge. This project emerges from a long-standing interest in the field that began with a monograph, *Political Anatomy of the Body: Medical Knowledge in Britain in the Twentieth Century*, published in 1983, and continued with a number of papers over the intervening years. As with the previous work, my debt to Michel Foucault's writings is large – though this time he is placed in historical context and therefore only appears as a recognisable figure/author in the final chapters. In some ways, this new book can be seen as a sequel to the earlier monograph in that themes and ideas recur; nevertheless, the overlap of empirical material between the two books is negligible and the ambition of the new is considerably greater.

Secondly, this book offers a medical history of the last 150 years. Given its relative brevity it is inevitably schematic and lacking in the detail that might be expected from traditional historiography. Further, the text also adapts some of the conventional customs of historical scholarship. The need to explain Man as the outcome of knowledge and practices means that the usual prioritising of the person (who 'knows' and 'acts') or the social group as the agent of history has had to be reversed. In effect, the narrative requires a history without actors or agents, that tries to reconstruct what exactly could be perceived at different times in the last hundred years or so through the anonymous eye of medicine. This implies an emphasis on text rather than authorship. To this end, the analysis also deliberately and exclusively uses primary sources. Of course, the story has been influenced by other texts that might better be described as secondary sources but to allow medicine to tell its own story these sources have been omitted except when they too can be construed as primary.

These theoretical and methodological issues are an important component of the book and what it is trying to achieve. There is therefore a case for setting out these arguments more fully in the opening chapter. Yet the book also presents a story and a discussion of method at the beginning would sit uneasily with the flow of the narrative. The decision was therefore taken to place these justifications and explanations in a final chapter (Chapter 17) – though this should not prevent the interested reader from starting at the end of the book before moving to the beginning.

The third and most ambitious aim of the book is to provide a creation story for Man. For well over a century now a Darwinian reading of nature has enabled us to understand where we came from and who we are. In place of Darwin's reading of nature, however, the approach adopted here is a reading of medical texts to report what medicine could see of the patient's nature and identity at any particular time. It is this changing object seen through the turning prism of medical perception that, it is argued, maps our mutating identity.

A problem encountered when writing this book was how to refer to the 'object' of medicine – the patient or person that clinicians deal with on an everyday basis. The identity of the patient is both multi-faceted (biological, psychological, social) and ever-changing so that it is difficult for any one descriptor ('person', 'patient', 'body', etc.) to capture fully the contemporary 'object' that medicine addressed at any given time without offering this object a sense of permanence. The solution adopted is to use 'identity' in the main title and in the second half of the book, which deals with the latter part of the twentieth century and the increasing salience of psychosocial dimensions of patient-hood, but to use 'Man' as a generic term, particularly for the nineteenth century period. Admittedly, the term 'Man' has taken on another layer of (sexist) meaning in the late twentieth century but for the earlier period, it still captures that Biblical and anthropological sense of identity and differentness from Nature.

Many colleagues have contributed to this thesis over the years, most unwittingly. To them my gratitude. I would like to thank Matthew Gothill for allowing me to develop material published in a joint paper as the basis for Chapter 15. Finally, I would like to acknowledge Jane Ogden for her discussion, support and encouragement during the writing of this book.

DAVID ARMSTRONG
London

1
Prologue

For many centuries, the Bible provided Western societies with their story of how Man was created. Throughout that time, it was incontrovertibly true that Man was created on the sixth day, between the making of the world and God's day of rest. In the mid-nineteenth century, however, this accepted account of creation was usurped with the publication of Darwin's *On the Origin of Species by Natural Selection* in 1859 and *The Descent of Man* in 1871. More than simply presenting a novel scientific theory, these two books laid the foundations for a completely new story of creation in which Man had his beginnings in a long-distant evolutionary past.

From its halting start, Darwin's new creation story rapidly achieved widespread acceptance, establishing a new orthodoxy about the origins of Man. Science replaced religion as the source of knowledge about Man's inception; the Judaeo-Christian Biblical story was reinterpreted as myth or allegory and creation stories embedded in every other religion were forced to yield to a new all-embracing explanation of beginnings. The few remaining adherents of alternative views of creation were marginalized as cranks and religious fundamentalists while evolutionary explanations went from strength to strength. Not only Man's origins but also his anatomy and behaviour, his emotions and attitudes, his susceptibilities and illnesses, could all be understood in terms of an evolutionary perspective that emphasized a genetic foundation and a special place in the natural order of things. Indeed, in terms of the breadth of its claims and its dissemination into all aspects of modern life, evolutionary theory transcends any religious or ideological movement of the past and might fairly claim to represent the most successful system of thought that has ever existed.

Ironically, the near total triumph of Darwinism over dogmatic religion served to replace one belief system with another, albeit one based on science rather than providence. Now, in its turn, this new set of beliefs about the origins of Man seems unassailable. Who can dispute the evidence that lies all around, from the Galapagos Islands to the African Rift valley, from the everyday world of biology to the complexities of the human genome? How can the sheer weight of articles and books that have offered argument and support for the truth of Darwin's story be challenged? But then there was once a time when the evidence of God's work seemed to act as unassailable proof of the wonders of creation.

Accounts of Man's origins – whatever their provenance – share a number of different elements. There has to be a world before Man that might be described as 'nature', and there has to be a moment at which Man's body materialized. Then, there may be a period during which his identity or mental faculties changed or evolved before he is recognisably 'modern Man'. In addition, of course, there has to be an accepted body of 'evidence' to support the plausibility of the account.

The evidence for both Biblical and Darwinian stories of creation can be seen as being embedded in texts. A text in this sense does not necessarily mean typeface on a printed page; text is anything that, given knowledge of the appropriate language, can be read. Thus, with knowledge of God's plenitude, of his great chain of being, it was possible to read the Biblical creation story in nature and in Man. In the sky, on the earth and in the sea, the book of creation lay open for inspection by anyone who had the language to read, interpret and praise. Similarly, Darwin provided a language for reading nature. Natural variation could be read in a different way, not in terms of God's omniscience but of mutation. Palaeontology could then read the fossil evidence as a form of text, each rock stratum another page in the book of creation, just as genetics read the genome as a text of evolutionary development. This new way of reading nature proved the crucial innovation that allowed Darwin to usurp the long reign of the Biblical creation story. Similarly, any challenge to Darwin is unlikely to come from a direct challenge to his theory but from the advancement of a new way of reading nature and the evidence surrounding the birth of Man.

The aim of this book is to show how a different story of creation, of the origins of Man, can be fashioned from 'evidence' that lies all around. Like its predecessors, this account will show how Man's body emerged from nature, how it came to be animated, and how a psycho-social identity was later fashioned. Like other accounts, it will also use the reading of texts as the building blocks for a narrative of Man's beginnings, in this case written texts, writings that can be used as a royal route to that preconscious point of origin.

Which texts, then, can provide a different reading of the evolving body and identity of Man? No doubt, the story lies embedded in many registers but medicine has particular claim to offer access to the key formative processes. In its search for illness and malfunction, medicine has always been intimately concerned with understanding the nature of Man. A history of medicine can therefore provide evidence – equivalent to 'God's work' or of the 'fossil' record – in the form of the textual residues left by the beginnings of Man and his subsequent development. By examining medical writings that describe the contemporary body through clinical perception it will be possible to trace the gradual revelation of Man. What sort of body did the doctor see? How was it to be interrogated? What was the nature of the patient who lurked behind the corporal presence? It is through the great eye of medicine, which on countless occasions explored, analysed and dissected Man, that the wellhead of Man's existence can be located.

There are many texts, most mundane and routine, that provide a view of Man refracted through the lens of medicine: descriptions of diseases and their manifestations intended for students of medicine, case histories in medical records, lists of symptoms in guides to the differential diagnosis, regimens for maintaining good bodily health in literature on hygiene, case reports of madness in psychiatric journals, and so on. Through these writings, it is possible to reconstruct what a contemporary doctor was both able to see and understand about the nature of the patient. How did the doctor perceive the object that lay quietly breathing at the end of the stethoscope? Was it in terms of humours or of electromagnetic forces or of cells, tissues and organs? The very act of applying the stethoscope and reporting the results – heart sounds, valve murmurs, and the like – gives a clue as to how the body was perceived at that particular time. Equally, other questions that medicine addressed such as:

'What motivated this patient who failed so frequently to take medication?' or: 'What were the characteristics of patients who spread diseases?' can be used to draw a vivid picture of the contemporary nature of Man and his attributes. The strategy is first to place these perceptions into some sort of chronological order, then to look for patterns, and so begin to develop the language that will allow these various medical texts to yield a coherent narrative of Man's origins.

So when should the story start? When does this new point of origin appear? The Bible started near the beginning of time, on the sixth day; Darwin and his successors identified a period several hundred thousand years ago when Man evolved from ancestors that were more 'primitive'. The narrative of creation described here, however, provides Man with a much more recent history that is less than two centuries old. Indeed, in this account, Darwin himself becomes part of the fossil record: *Origin of Species* and *The Descent of Man* are thereby transformed from the explanation to what needs to be explained. The nineteenth century Darwinian revolution marks not the end of the search for origins but the moment of its beginning.

2
Constructing the Body

In the beginning there was no Man. Disease was not mapped onto human anatomy as there was no body on which to inscribe the contours of illness; medicine addressed a world without boundaries, a primaeval space characterized by shifting humours and movements of cold and damp. Early nineteenth century accounts of public health show that the primordial landscape in which Man was to crystallize in modern form was a world governed by natural forces. 'Elemental disturbance (such as) long drought, excessive heats, hot burning winds, clouds of suffocating dust, heavy rains ...' (Bascombe 1851: 189) provided the environment in which epidemic pestilence could emerge. The origins of many diseases were grounded in meteoric phenomena (Haviland 1855) or in low, marshy and alluvial soils (Parkin 1859). This was an inhospitable world, of earth, sky, and weather, in which diseases roamed free.

The little segmentation that did exist in this open landscape was based on the lines of separation drawn by rules of quarantine. Quarantine, first used in the Middle Ages as a means of isolating ships believed to be carrying diseases, operated by restricting movement between defined spaces, say, between the space of a ship and that of the land, or between one ship and another. This strategy of segregating one space from an adjacent one by drawing and maintaining a *cordon sanitaire* gradually spread as lines were sporadically drawn around houses, streets, towns and countries whenever transmission of disease threatened. Each line served both to prevent movement between spaces and to delineate the separated spaces across which movement was forbidden.

In England the last attempt at extensive quarantine occurred in 1831 with the outbreak of Asiatic cholera. A board of eminent medical men was assembled at the College of Physicians under the chairmanship of Sir Henry Halford. From their deliberations, they issued an official notice that was published throughout the country in the form of an Order of the King in Council. The sick, they pronounced, should be kept separate from the healthy, all being moved to a designated place in each town or its neighbourhood:

> And in case of refusal, a conspicuous mark, 'SICK' should be placed in front of the house, to warn people that it is in quarantine; and even when persons with the disease shall have been removed, and the house shall have been purified, the word 'CAUTION' should be substituted, as denoting suspicion of disease; and the inhabitants of such a house should not be at liberty to move out or communicate with other persons until, by the authority of the local board, the mark shall have been removed.
>
> (Quoted in Smith 1866: 62)

Some of Halford's extreme recommendations were not adopted though the intercourse between one town and another by sea was prohibited. Nevertheless, the approach illustrated the continuing importance in early nineteenth century public health of the principle of preventing passage between defined localities. Quarantine marked a line of exclusion between places that could not be transgressed: country might be separated from country, town from town, house from house. Indeed, Halford and his colleague advised that:

> it may become necessary to draw troops or a strong body of police around infected places so as utterly to exclude the inhabitants from all intercourse with the country.
>
> (Quoted in Smith 1866: 62)

Under a system of quarantine, illness somehow resided in places as it was places that had to be kept separate. To be sure, the movement of people was controlled, but only because they were the means through which one place was brought into contact with another. Even when the ill and dying had been removed from a house, the place itself remained suspect and continued to be subject to quarantine regulations. There was no perception of microbial infection, as there would be later in the century; danger lay in a defined geographical space and its ability to communicate with other localities. The public health problem was the uncontrolled movement through which such dangerous unions could be made. This movement could not be arrested except by total exclusion, by ensuring that the connections between one place and another were systematically blocked. Policies of quarantine therefore operated in an open landscape, across a geographical terrain, marking and policing lines of separation between one locality and another. This was a geographical system of control in which Man, as an individualized figure, did not appear; nevertheless, these several basic elements – landscape, movement, and lines of exclusion and separation – provided the formative conditions and the embryonic space in which the body of Man could materialize.

Sanitary science

Writing in 1866, Southwood Smith, the famous pioneer of modern public health in Britain, was dismissive about the 'obsolete doctrines' of public health that underpinned a belief in quarantine:

> True safeguards against pestilential diseases are not quarantine regulations, but sanitary measures ...
>
> (Smith 1866: 64)

Sanitary science at once gave a name to the former anonymous movement that brought one space into contact with another: the new danger was dirt. But the great innovation of sanitary science was to interpolate a new space of hygiene into the more traditional geographical map of public health, namely that of the volume of the human body. Instead of a *cordon sanitaire* between potentially coalescing geographical spaces the new regime of hygiene monitored a line of separation between the space of the body and that of its environment. At the same time, the internal characteristics of this corporal space were also being dissected and studied in the newly emergent medicine of the clinic and the hospital; yet it was public health that grappled with the fundamental question of body boundaries, of the lines that demarcated a corporal space from a non-corporal space, of the separateness and individuality of each human body that fell within the purview of the sanitary authorities.

Sanitary science addressed two major inter-related problems. First, by applying the old science of landscape to the contours of the human body and mapping its topographical relations a new question arose, namely what belonged to corporal space and what lay outside of it? Secondly, having defined this space, it was not simply a case of drawing and policing a *cordon sanitaire* to protect it. Air, water and food had to pass across the boundary into the body and, equally, waste products such as faeces, phlegm, sweat and urine had to be removed. This meant that a regime of complete exclusion as under quarantine was impossible. The focus of late nineteenth century public health therefore became the zone that separated anatomical from non-anatomical space, and its approach to hygiene involved the monitoring of matter that moved between these two spaces, especially in the form of dirt. As Fox, in his handbook of 1878 for Medical Officers of Health, noted:

> The elementary principles on which the greater part of the Medical Officer of Health is based, may be truly said to be the prevention of the pollution of Water and Air with filth and its products, and the prevention of the consumption of particles of food deleterious to health.
>
> (Fox 1878: 1)

In part, the strictly sanitized passage of substances between the outside and the inside of the body embraced the older environmental preoccupations with soil, climate and buildings but now only in as much they could be involved in the contamination of bodies. Soil, climate and buildings were simply the intermediaries between bodies and the hazards that threatened them; the main goal of the new sanitary hygiene was to establish a network of surveillance over the entire 'natural history' of those substances that might cross the boundary of corporal space – air, water and food passing inwards, and body wastes and secretions passing outwards.

Air moved from its former status as a constituent of atmosphere to a potentially hazardous substance. It could be contaminated by suspended matter, it could contain noxious fumes and gases, it could be characterized by shortages of oxygen or excesses of carbon dioxide. To these dangers could be added impurities from combustion and air vitiated by effluvia from sewage matter. The problem was even more acute in poorly ventilated rooms due to the:

> presence of scaly epithelium, single and tessellated; round cells like nuclei, portions of fibres (cotton, linen, wool), portions of food, bits of human hair, wood and coal ... statistical inquiries on mortality prove beyond doubt that of the causes of death, which usually are in action, impurity of the air is the most important.
>
> (Parkes 1873: 90)

'Our greatest enemies', declared de Chaumont, 'are *foul air* and *foul water*' (De Chaumont undated: 5). Sanitary science concerned itself with calculating precisely the body's needs for water (a minimum of 12 gallons of water daily for each person for domestic purposes). In addition, considerable amounts of water were needed for the operation of water closets and for keeping animals. Provision of such quantities of water called for elaborate schemes of collection, storage and distribution, all of which involved risks from impurities. In public health terms, the supply of water therefore carried the twin dangers of insufficiency and contamination:

> the person and clothes are not washed, or are washed repeatedly in the same water; cooking water is used scantily, or more than once; habitations become dirty, streets are not cleaned, sewers become clogged; and in these various ways a want of water produces uncleanliness in the very air itself.
>
> (Parkes 1873: 35)

Like water, food could also be contaminated with dirt during its movement from the land to the mouth. In part, the 'civilizing processes' of eating customs and table manners were recruited to the task of ensuring food purity, but the responsibility of the sanitary authorities ranged across the whole chain of food consumption and production, extending from the farm to the shop to the kitchen to the table. Food inspection to ensure the exclusion of contaminants was a major part of sanitary work, nowhere more important than in the inspection of animals' bodies. At the moment when the human body was being so carefully demarcated the threat posed by another 'body' was a particularly potent one; the divide between the 'natural' body of the animal and the new space of human anatomy was therefore managed with especially rigorous surveillance techniques.

Ensuring purity of air, of food and of water was only half the task, and the easier one at that. Substances entering the body could be monitored throughout their natural history, but the discharge of bodily effluent posed problems of a different order. The latter was particularly hazardous because, for a brief moment, it bridged corporal and non-corporal space and then needed urgent decontamination. Removing waste substances from the environs of the body was only the first step before the danger was finally extinguished by transforming unsafe bodily products into non-corporal neutralized matter. Yet this process of decontamination was a difficult one. Exhaled air could be removed by diffusion, convection or artificial ventilation to be returned to the world of atmosphere and climate, while sweat, urine and faeces could be washed away to the soil and sea. These methods, however, relied on a process of dilution that meant that at any point in the process of degradation impure matter, in its turn, could pollute the environment with which it was meant to be reintegrated. For example, the sanitary disposal of sewer water posed major problems in terms of the best methods of decontamination. Was it to be fed into rivers, or the sea, or into tanks with overflow, or purified by precipitation, by filtration, or by irrigation?:

> At the present moment the disposal of sewer water is the sanitary problem of the day.
>
> (Parkes 1873: 351)

In effect, besides ensuring the safe transfer of excreted substances across the body boundary, sanitary science had to oversee the efficient functioning of an active process of degradation and transformation. Dirt from the environment entering the body was dangerous; yet even more hazardous was the potential threat from excretory products coming back across the body boundary to cause harm. The products of the body – sweat, exhaled air, urine, phlegm, vomit, faeces – took on a new abhorrence: excretions and secretions of the body posed the greatest danger to health and elicited the strongest hygienic response.

The one apparent exception to the obvious inherent dangers of body excretions was semen. Certainly, it emanated from the body but it was hardly a waste product. Nonetheless, ejaculation of semen held its own distinctive dangers, especially in the form of masturbation. Nineteenth century public health gave masturbation, as a form of 'unnatural' ejaculation, a special place in terms of the almost unspeakable dangers it posed to health and happiness:

> The drain on the system produced by solitary vice, arrests development to a considerable extent, and prevents the attainment of the strength and endurance which would ensure a healthy, vigorous and happy life.
>
> (de Chaumont 1887: 376)

Contemporary concerns with dirt, cleanliness and hygiene manifested themselves in writings on the dangers of specific substances. At times, the texts exploring these public health dangers barely mentioned the body or body boundaries, yet the ghostly presence of the body was never far away. Even if sanitary science as a whole had never explicitly addressed or defined a body boundary, it would still be possible to see the contours of corporal space writ large in nineteenth century public health. The 'new' concerns with substances passing into and out of the body spoke loudly about the existence of an anatomical space separated from its outside in a way that was entirely alien to the former public health regime of quarantine. The very fact – for so it had become – that there was an inside and an outside meant that there had to be a line of separation between the two, a border that defined the limits of a body for Man.

Body boundaries

As sanitary science attended to the controlled passage of substances between the space of the body and non-corporal external space, it inevitably implied an interface across which movement occurred. This interface was, or more correctly, would become, the body's boundary. That is why, of all bodily organs, systems and tissues, sanitary science became particularly concerned with the hygiene of the skin, the mouth and the bowels.

The skin marked the most obvious boundary between corporal and non-corporal space; yet more, it was viewed as an extended organ of excretion and therefore the target of sanitary practices. The skin discharged waste materials from the body's interior to its exterior and these had to be removed. The skin therefore had to be regularly washed, disinfected and deodorized. Moreover, it had to be maintained in health through being cooled or kept warm, exposed or clothed. Inadequately maintained it was liable to skin diseases such as 'scabies and the epiphytic affections especially' (Parkes 1873: 35). Even so, the boundary of the skin was relatively clear-cut, and other than in its capacity to sweat, remained of relatively minor sanitary concern; much more dangerous were those bodily places such as the mouth and rectum across which dangerous substances had to pass and which ambiguously bridged internal and external space.

The mouth was a hazardous interface because it was the first point of contact for many extraneous substances:

> It is reasonable to suppose that (impurities) would be likely to produce their greatest effect upon the membrane with which they come first in contact. This is in fact found to be the case.
> (Parkes 1873: 38)

Moreover, it was not clear whether those substances in the mouth had actually crossed the body's boundary. Was food taken into the mouth now in the body? Or was it still outside? What was clear was that at the point it entered the mouth food was rapidly transformed from a wholesome object into a potentially dangerous one. Half-chewed food could no longer be shared just as excretions from the nose or spittle from the mouth could no longer be discharged casually into the outside world. Substances in the mouth, the mouth itself, and the surrounding nasal orifices therefore demanded the closest scrutiny by the sanitary authorities, particularly in terms of elaborate regimes of oral hygiene and emphasis on the health of the teeth:

> The mouth and all mucous orifices should be kept scrupulously clean ... There can be no doubt that by keeping the mouth thoroughly sweet and clean and by stopping carious teeth as soon as discovered (a person's) vitality may be greatly prolonged.
> (Newsholme 1892: 340)

Food entered the body, was processed and absorbed, and waste matter was discarded. Before it was discarded, however, it was actually stored in the space of the body. Faeces were therefore doubly dangerous. On the one hand, they needed rapid removal and decontamination once voided; this meant that an important responsibility of sanitary science was 'to remove as rapidly as possible all excreta from dwellings' (Parkes 1873: 336). On the other hand, and of even greater sanitary concern, faeces needed excreting as regularly as possible – perhaps followed by colonic irrigation to ensure complete cleanliness. The problem of constipation ('allowing food to remain even to decomposition, as leading to distention and sacculation of the colon, and to haemorrhoids' (de Chaumont 1887: 376)) became an important public health concern in the late nineteenth century and, in consequence, a regular bowel movement became a part of a healthy sanitary regimen.

Taken together, these various processes of monitoring and defining the body boundaries represented a marked change from the *cordon sanitaire* of quarantine. Under a system of quarantine, illness had belonged to places: it was places that had to be separated from places because it was the place that harboured the disease not the person. During the cholera epidemic of 1831, for example, food for the ill had to be placed in front of the house and received by one of the inhabitants of the house after the person delivering it had retired. The problem was not that person could infect person directly but because person carried the taint of place that would allow contiguous but separate areas to be dangerously mixed. A house vacated by the ill and dying remained suspect; conversely a person who happened to be within a place that was judged dangerous automatically fell within the embrace of quarantine regulations: yet another anonymous body caught within a space defined by the lines of exclusion. Quarantine only conceptualized the person as the nameless figure that might break through the essentially geographic separators that held illness at bay.

Quarantine had operated in a landscape bereft of Man. The landscape was a 'natural' phenomenon, an assembly of continuous geographical structures that began to be divided one from another by ephemeral lines of separation, spaces being sub-dividing into spaces. It was a world of dust and dirt, but one from which sanitary science, reconfiguring the old explanatory framework of topography, dirt, lines and movement, began to separate out the new space of an individualized anatomy. The application of rules of hygiene involved a concept of cleanliness and purity, a separation between one space or object and another, so that mutual contamination did not occur. Quarantine marked out a geographical space while sanitary science defined the space of the body. And with the new lines of differentiation the world of elemental forces was largely tamed. By the end of the nineteenth century, even those great natural forces of the earth and the weather had been subsumed under sanitary science while natural matter such as soil became implicated in disease precisely because of its place in the web of sanitary interchange (Miers and Crosskey 1893). As Poore observed in his Milroy Lecture of 1899:

> That which we commonly speak of as earth, soil or humus, is largely composed of excreta and the dead remains of animals and vegetables, which, as the result of fresh biological processes, are either returned to the bodies of living vegetable organisms or ... find an exit in the sea ... the line of demarcation between earth on the one hand and water on the other is often not very definite.
>
> (Poore 1902: 3)

Where quarantine had policed only a boundary, sanitary science monitored a whole population. The ubiquitous presence of dirt allowed no one to be exempt from surveillance. This vision of a total public hygiene, as Guy observed in 1870, 'has to do with persons of every rank of both sexes, of every age'(p. 6). Moreover, it focused not only on their persons but also on their whole lives:

> It takes cognisance of the places and houses in which they live; of their occupations and modes of life; of the food they eat, the water they drink, the air they breathe; it follows the child to school, the labourer and artisan into the field, the mine, the factory, the workshop; the sick man into the hospital; the lunatic to the asylum; the thief to the prison.
>
> (Guy 1870: 6)

Sanitary science recognized a great cycle of interchange involving contamination and purification as substances passed between the anatomical space of the body and the geographical space that surrounded it. The practical regime of hygiene that was introduced to monitor this great exchange had to address constantly the separation of these two spaces, corporal and non-corporal, anatomical and environmental, watching the passage of substances across the body boundary, identifying the point of separation marked by the skin, but managing the problematic areas of colon and mouth with more intense scrutiny. The effect of this constant activity was to sculpt a new anatomical shape, giving it separation and substance. Older ideas of illnesses that moved through the body, within and without, providing no clear boundary between what was body and what was non-body, were superseded. The new regime of sanitary science made the body boundary into the target of everyday practices, cleansing and monitoring, sanitizing and surveying, that in its turn brought to the searching eye of medicine the firm outline of the corporal space of Man.

While the shape of the body of Man could largely only be inferred from the practical routines of sanitary science, its emerging morphology could be seen laid out graphically on the printed page in the first edition of Henry Gray's celebrated anatomical textbook, published in 1858. As was later to become so self-evident, the diagrams and text described a body made up of organs, tissues and cells. It was the routinisation of clinical examinations, dissections and post-mortems in the early nineteenth century that constructed this internal plan of the human body; it involved internalizing geography, so to speak, by applying the science of cartography in the construction of the human anatomical atlas. Thus, clinical medicine provided the internal map of the human body while sanitary science defined the boundaries within which it was contained.

But there was a time, not that long ago, when the body was not seen in this way, a time in which humours moved in what were seen, in retrospect, as mysterious ways, a time when the 'external' environment was as much a part of the illness experience as the marks of sickness in the body. Although not the first anatomical text, Gray's *Anatomy* can be used as a marker for the moment when Man's body appeared as separate, divisible, and analysable.

What was this body like? There is no answer beyond how it was perceived in the various medical texts. It could be perceived as a bounded corporal space in sanitary science or as a cross-sectional drawing in an anatomy book. This was no ready-made body waiting to be discovered by an enlightened medicine. This was no universal truth, long hidden but now revealed. This was a body that was crafted, whose dimensions and attributes were the outcome of painstaking work by anatomists and clinicians and public health officials; the body was no more and no less than what they could perceive and describe. In its essence, it was a bounded three-dimensional space separated from a world outside, and increasingly internally sub-divided. This space was the body, the body was this space; these lines of separation provided the volume in which the body could materialize as a solid living density.

In this history, the origins of Man as a discrete physical body do not lie in an African rift valley many thousands of years ago; indeed, it only became possible to search for such mythical origins after the body had been fabricated. Far from the romance of physical anthropology and palaeontology providing an account of Man's origins, these sciences of beginnings could not have existed prior to the first invention of Man. How could ancient rock formations be read to provide a history of Man before the body of Man had begun to materialize? Darwin can surely be placed alongside the pioneering clinicians and public health officials as one of the great progenitors of Man: while they, in their mundane routines, fabricated the body of Man, he devised for it an epic history of descent.

The argument of this new creation story, that the body of Man was invented in the mid nineteenth century, does not mean that there cannot be a fictitious line of descent. In 1818, Shelley published her novel *Frankenstein* (within two years of Laennec's account of his invention of the stethoscope, better to listen to the internal sounds of a three-dimensional body). This story of a body constructed of separate parts can stand as the representation of proto-Man, the immediate antecedent of Man, the primitive Neanderthal, the missing link between a swirling landscape and a newly crystallized corporal identity.

3
Negotiating Death

The outline of the new figure of Man was an artefact of those nineteenth century sciences that established a new corporal space separate and free from nature. Darwin's contemporary *Descent of Man* also provided a story of how Man came out of nature, but could not reflect on its own critical role in that process. Darwin recognized (as did his contemporary Marx) that the integrity of Man was based on the completeness of his emancipation from nature and, through the mechanism of evolution, described the forces that turned early Man into modern Man. Such a series of sequential steps to Man's history might serve well the fable of his origins, but in the nineteenth century, there was no logical ordering to Man's invention – the child followed the Man, birth followed life, life followed death. It was only later that these fragments could be assembled into a coherent narrative of beginnings.

Paradoxically, the end of Man provided the main point of articulation for an understanding of his beginnings. Death was the culmination of life, yet in a great reversal became the vantage point from which to construct that very existence; and ironically, clinical medicine with its declared purpose of keeping death at bay actually introduced death into the core of the body. The new clinical medicine of the nineteenth century re-conceptualized the meaning and significance of death, locating it in the cells, tissues and organs of the body's interior, and at the same time re-ordered the practical and sanitary procedures that surrounded the lifeless corpse. In so doing, the body of Man was further prized away from nature as the new figure of death began to breathe life into the body's quiescent space of existence.

Pathological death

In the eighteenth century, death was a dramatic figure in a black cloak, scythe in hand, who knocked on the door of life. Death had come from outside of life: sometimes it could be resisted, forced to make its visit another time, but eventually it gained entrance. Then, during the early decades of the nineteenth century, a new model of illness that would dominate clinical practice until the closing decades of the following century replaced the older theories. The new model of medicine localized illness to a specific pathological lesion inside the body. Disease was not a fleeting constellation of symptoms moving in and out of the body; disease was an abnormality of structure or function contained within the body. One possible consequence of this internal abnormality was such disruption to the basic functions of the body's systems as to cause death.

Built around the formative idea of the pathological lesion, the new medicine destroyed the age-old figure of death. It was not death calling from outside that ended life but the effects of the pathological lesion inside the body. The seeds of death were enclosed in the body in the same way as were the springs of life; death did not come at the end of life but was contained in the body alongside life; the body was dying even as it was being born; cells began their dying trajectory at the moment of procreation. As William Farr, the first medical statistician at the British General Register office, noted in describing the workings of the analysis of cause of death in the new registration procedures, 'the human body has a tendency to death; but the tendency to life is stronger in almost every instant of existence' (Farr 1839: 89). In consequence death could be individualized: instead of a generic death from 'natural causes' for almost everyone (except when the coroner decided 'unnatural' events had intervened) a specific label for the proximate pathological 'cause of death' could be ascribed each and every body.

The notional existence of a specific cause for every death meant that each body could be explored after death for the pathological lesion – the secret of death – that had caused life to be ended. It was also possible to devise an elaborate system of classification and collation that would transform death from a private event to a public statistic. In Britain, for centuries, deaths had been recorded in the parish register: the deceased's name, sex, age and profession or calling. With the civil registration of death, however, a new analysis was instituted that required an additional item of information – the cause of death (in the form of the disease that had apparently brought about death). These entries were, in their turn, transferred to central registers, collated, analysed and published.

The new medical analysis of the cause of death gradually began to usurp other rituals for managing the transition between life and death. In place of the tolling of the church bell, the religious procession carrying the Corpus Christi, the friends and relatives clustered around the bed in the darkened room, there was a new ceremonial marked by the mundane completion of the death certificate. A natural death was a domestic experience, set before the family and neighbours. The new death involved clinicians, pathologists, coroners, clerks and registrars who subjected the corpse to a detailed scrutiny to establish the true cause of death. Medicine ushered in a new regime of investigation and analysis around the body that did not search for the familial bonds of the dead for a mirror to the truth of life, but instead examined the internal organs of the body itself where both the core of life and death reposed.

The medical revolution that ushered in pathological death was more than a new way of thinking about illness. Medicine transformed death, and with it redefined life. Life was both the natural force contained within all living things and the corporal energy that grappled with death in uneasy equilibrium until it was finally overcome. It was not life being overwhelmed by an outside death but an inner death that called forth life to resist it. Corporal space, delineated by sanitary science and sub-divided by human anatomy, was imbued with a life force pulsating through its inner tissues and organs: Man had corporal life because he had pathological death lurking in the heart of his body. No wonder that the truth of life was now to be found in death as the pathologist dissected the body, the clinician completed the death certificate, and the registrar collated these records of the epic struggle within the body of Man.

The new pathological death served to remove Man further from nature. Man was no longer a part of the natural world and subject to those forces that produced a natural death but a part of a separate and independent domain that was established through his inner death. In the world of nature 'natural deaths' reigned; in the world of Man pathological deaths acclaimed Man's differentness. During life, the clinician searched the body for the pathological lesion and after life, the pathologist opened up the corpse in the post-mortem or autopsy to reveal the truth of death in the form of diseased processes or structures that had brought it about. Both of these examinations of the patient's body served to reaffirm, on countless occasions, its three-dimensional volume as corporal space was mapped, analysed and dissected.

The clinical examination and the post-mortem were the major practical procedures for identifying and managing the new pathological death. Yet, when the clinicians and pathologists, the coroners and registrars, had all had their say and the body had finally yielded its secrets, there was one final problem. How was the lifeless body to be disposed of? In its entirety, the process was a great cycle of dust to dust, ashes to ashes, dirt to dirt, but the flickering image of the body had appeared in the gap that separated one region of dirt from another. Sanitary science had ensured that the flame burned increasingly bright in its own exclusive space methodically demarcated from its surroundings; but after death the flame was extinguished, and corporal space became an empty hollow that had to be returned to dust.

Disposal of the dead

How to dispose of the dead was the most vexatious problem faced by sanitary science in the nineteenth century, yet was, in a way, one of its own making. While living, the anatomical space of the body was maintained in its separateness by elaborate sanitary procedures, but when dead, the body itself had to be rejoined to non-corporal space. This process posed major difficulties whichever way it was perceived. Sanitary science had split the world into two parts, nature and a body-separated-out, but then, at the end of life, those worlds had to be collapsed together again, the crucial distinction between anatomical and non-anatomical space dissolved. Such a process threatened the inviolability of the rules of hygiene that underpinned sanitary practice. Sanitary science had maintained the integrity of the body by trying to monitor and exclude these dangerous substances – mainly dirt – from crossing the great divide: how then could the body, this great creation of hygienic practice, be reduced to something akin to dirt? If the corpse was construed as another form of dirt then surely it posed even greater dangers when it crossed the now hazy line between body and non-body. So much energy had gone into differentiating the body from its surroundings but now these efforts had to be undone and the process reversed. What had been made sacred had to be made profane, and sanctification was always an easier process than de-sanctification.

Disposal of the dead had previously belonged to the domain of religious practice, but with the newly realized public health dangers, the corpse became an object of sanitary law. It was not a case of the Church sanctioning the separation of the soul from the body but the sanitary authorities releasing the body across the great divide once more to become non-body, to decompose into dust and dirt. The old regime had allowed the body, that empty husk of life, to be dumped with general indifference into the earth whereas the new public health closely regulated the progression of the dead from the world of the living to the world of nature. There could be few objects more dangerous to the health of the population than the decomposing corpse and until it had made the transition back to nature, until it had fully departed from the world of corporal space to which it had once belonged, the sanitary authorities and the public had to be ever vigilant:

> One of the most warmly contested questions in the field of sanitary reform which has attracted public attention during recent years had been the disposal of the dead.
>
> (Wilson 1892: 535).

There was the immediate problem of the relatives delaying the disposal of the dead body through an:

> unreasoned sentiment which prompts the retention of the body in overcrowded homes for as long a period as possible... . if a dead body is exposed to a temperature of 60 Fahr., it will begin to putrefy in three days, and give off offensive gases, and numerous cases of illness have been attributed to this cause alone, apart altogether from specific infection.
>
> (Wilson 1892: 563)

Even so, the focus of sanitary measures was the series of techniques through which the corpse was transposed from a dangerous object to a safe one. Burial had been the customary way of disposing of the dead but such a procedure could well be insanitary, particularly 'the baleful effects of the practice of interring the dead in the midst of the living' (Greene 1857: 1) when burials occurred within towns. Burial places became overcrowded, the ground often rising above its original level, and graves were only partly filled in (in expectation of the need to accommodate other family members); everywhere smells and nausea, and 'constantly the dreadful effluvia of human putrefaction' (Walker 1839: 9).

Towards the end of the century, cremation became more popular, promoted by hygienists. Some public health physicians, however, opposed it:

> It is true that the impurities in burning can be well diffused into the atmosphere at large... . But if the burning is not complete, foetid organic matters are given off, which hang cloud-like in the air, and may be perceptible, and even hurtful.
>
> (Parkes 1973: 441)

There was also burial at sea but this too could result in the body or semi-decomposed body parts being washed ashore. Nevertheless, because of the more immediate and obvious dangers, burial in the ground remained the greatest challenge to sanitary science:

> Burying in the ground appears certainly the most insanitary plan of the three methods. The air over cemeteries is constantly contaminated, and water (which may be used for drinking) is often highly impure.
>
> (Parkes 1873: 441)

A Royal Commission had been established in Britain in 1850 to report on the question of burial and a succession of Burial Acts followed (Greene 1857). Regulations were introduced to ensure that new burial grounds were sufficiently distant from towns and individual dwelling houses; walls or railings were required to be at least eight feet high; plot size was stipulated and depth of burial specified. The Disused Burial Grounds Act of 1884 allowed no building to be erected on any disused burial ground. The 1863 Regulations under the Burial Act permitted no grave to be reopened within 14 years after the burial of a person above twelve years of age (eight years for younger children) unless to bury another family member:

> in which case a layer of earth not less than a foot thick shall be left undisturbed above the previously buried coffin.
>
> (Hamer 1902: 558)

After death, the buried body returned to its elements but this could be a slow process. Fourteen years was the minimum time required for a body to decompose but even then it was felt that gaseous and volatile substances and liquid products of decomposition might still contaminate the ground:

> In some soils the decomposition of bodies is very slow and it is many years before the risk of impurities passing into air and water is removed.
>
> (Parkes 1873: 440)

This meant that burials underneath or within the walls of any church were prohibited; burial in vaults or walled graves was also forbidden unless the coffin was 'separately entombed in an air-tight manner' and it was thereafter 'never disturbed' (Hamer 1902: 558). Judicious siting of the cemetery, good depth of burial and the appropriate use of plants could aid the decomposition of the corpse:

> deep burial and the use of plants, closely placed in the cemetery. There is no plan which is more efficacious for the absorption of the organic substances, and perhaps of the carbonic acid, than plants .. the object should be to get the most rapidly growing trees and shrubs.
>
> (Parkes 1873: 441)

Even with these precautions, the corpse remained a hazardous object for many years and the process of decomposition was an uncertain one. Hamer cautioned that if on re-opening any grave:

> the soil is found to be offensive, such soil shall not be disturbed and in no case shall human remains be removed from the grave.
>
> (Hamer 1902: 559)

Cemeteries were always potentially dangerous places with noxious odours, poisoned run-off waters and the perpetual risk that an undecomposed corpse might break through to the surface. These images belong to the routines of sanitary science but also to the horrors of the gothic novel. Indeed, the 'Dracula' genre of nineteenth century tales of the macabre from the likes of Poe and Stoker provide a graphic indication of the dangers of corpses and burials. Recurrent themes were the dangers arising from premature burial and the problem of the 'undead'. Both of these concerns reflected the potential horrors that were held to lie in the space between the living body and the decayed cadaver that had returned to the earth. But the undead were not a problem waiting to be solved by a Van Helsing-like hero; the undead were a product of a new space that opened up in the mid nineteenth century, a void that continues to resonate its unsettled memory. The undead marked that dangerous space, that hazardous transition, between corporal and non-corporal worlds, a space for both sanitary practice and gothic tales, but also a space that served to reaffirm once again the nascent contours of a materializing Man. It might be said that the body of Man made its appearance between Shelley's *Frankenstein* of 1818 and Stoker's *Dracula* of 1897, between the horror of a body that was constructed from separate parts then resurrected with the energy of life to a one in which life and death co-existed in unwholesome fellowship.

4 Discovering Origins

The new pathological analysis of death consolidated the separateness of Man's body from nature and infused him with an intra-corporal life. Yet the fertility of death was not yet done. Death had constructed the end of life; it was now turn to discover its beginning.

The rituals surrounding the disposal of the corpse had been directed at negotiating the great divide between corporal and non-corporal space. At the other end of life, the problem involved the reverse process: how to secure a separation of Man from nature at the moment of birth. Just as the Church had for centuries processed bread into the body of Christ, so the medical authorities had to manage a more profane process of transubstantiation through which the body of Man emerged phoenix-like from dirt. Yet too often, the infant was perceived as an object that could so easily slip away back into nature through the processes of atrophy and decay. The challenge was to replace the natural processes of degradation and involution that threatened infancy with ones that respected the qualitative distinction between body and non-body. The project was to identify the beginning of life, that moment when Man was realized, and then to ensure that he emerged fully from the canvas of nature. The task was to establish a point of origin for Man.

The origin of life

In 1857, the British Registrar-General published his customary Annual Report in which, for the first time, he described the deaths of a new object, the human infant (Registrar-General 1857). He had analysed deaths by age since his third Annual Report published in 1841 but this was only for three large cities and laid no particular emphasis on deaths amongst infants. The 1857 figures however provided the number of deaths under the age of one year for the whole country and in subsequent years he published an annual breakdown of infant deaths by various causes. In 1877, this analysis was formalized in the construction of a new statistic, the infant mortality rate.

The calculation of the infant mortality rate involved dividing the number of deaths in the first year of life by the number of all live births. The number of deaths was easily established but the denominator, the number of live births, was more difficult to establish. What exactly was a live birth? For the Registrar-General the defining characteristic was whether the new-borne infant took a breath: the demarcation line between existence and non-existence was a single gasp for air. This meant that within the Registrar-General's classification of the late nineteenth century, a stillbirth at full gestation was indistinguishable from an earlier miscarriage or an abortion as none of these infants ever exhibited the vital spark of life. Birth and a breath marked the beginning of existence.

The emphasis on a life beginning at birth was furthered by the recognition of prematurity in 1881 as a distinct form of developmental disease that could lead to death. Thus, an infant born prematurely that took one breath was an infant who lived then died, whereas an infant that was delivered at term and failed to breathe was not considered to have ever had separate existence. Life was defined by a moment of independence so that even deaths from prematurity, which could so easily have been seen as a problem of the mother, were viewed more often as 'diseases of the child which thus asserts its autonomy' (Registrar-General 1870: 200).

The story of origins that located Man's beginning in an initial if brief sign of life lasted for about 50 years. By that time, it was becoming increasingly apparent that the use of a single breath to mark the difference between life and non-life was too imprecise. Who could tell whether a new-born infant took a breath of air and achieved momentary independence or made a reflex gasp before slipping back to nature? The solution was to revise the basis of the classification system from an emphasis on the physiological response of the infant to one that stressed gestational age. Inevitably, this involved concurrently revising the meaning of a stillbirth.

In 1926, the Births and Deaths Registration Act defined a stillbirth as occurring after 28 weeks of pregnancy. This meant that stillbirths could be separated from miscarriage and abortions, which now occurred, by definition, before 28 weeks. Infant life now possessed a formal beginning that occurred after a defined period of gestation in the womb. The difference between a live birth and a stillbirth was therefore no longer of significance in determining the origins of life and the distinction was progressively blurred. Instead of stillbirths being linked to all of infant mortality (in the first year of life) they became linked only to the first month: stillbirth and neonatal mortality (deaths in the first month of life), the Registrar-General noted, were 'closely allied' (Registrar-General 1938). A further refinement occurred in 1948 with the construction of the perinatal mortality statistic that measured deaths in the first week of life plus stillbirths with a denominator of all births and stillbirths. With that extension of the classification system the nineteenth century role of stillbirth as providing the link between nature – and non-existence – was finally dissolved.

In the early twentieth century the problem of deaths from 'premature birth' also began to disappear as the term no longer marked the process of nature dragging back the momentarily independent infant, but simply described the birth of an infant before full gestation. The introduction of a new death certificate in 1927 allowed some choice by the certifying doctor as to the cause of death; the result was a rapid decline in deaths from premature births. It had seemed, noted the Registrar-General writing in 1947, that the use of fixed rules of selection had given 'undue weight to prematurity as against certifier's preferences' (Registrar-General 1947: 31). The term 'immaturity' replaced 'prematurity' in the sixth Revision of the International Classification of Disease of 1950. This new designation enabled the infant to be classified twice, once with a specific pathological cause and with the label of immaturity as a context of death.

The new concept of immaturity shifted the old prematurity from an affliction of the child to a problem of the mother. It was premature labour that caused immaturity so immaturity represented pregnancy failure, a pathology of the mother not of the child. This meant that immaturity could not be used in all accuracy as a specific cause of death of infants in the classification tables; at best, it signified an interaction between mother and child but it no longer belonged to the proper analysis of infant mortality:

> Infant mortality would be more precise if it were possible to show separately the death risks in respect of infants successfully carried to term and infants that fail to reach term as judged by a simple criterion of maturity such as the infant's birth weight, or the length of the gestation.
>
> (Registrar-General 1954: 30)

By the mid twentieth century stillbirths, perinatal, neo-natal and infant deaths – together with their various derivative statistics – mapped out for medicine the first year of life. In its temporal dimensions the classification celebrated the moment of separation when a natural object was transposed into human life. It had been a 70-year journey from the first formal recognition of infant deaths to a strict definition of the beginning of life, and then a further 30 years to refine the boundary between life and death. To be sure, the Adam-like figure of the human body was fully formed by the close of the nineteenth century but the narrative of ancestry had to wait until the middle of the twentieth century to be finally completed. At last, Man had a point of origin to add to a demarcated body, as well as the necessary *rites de passage* after death to transfer this object back to the world of dust and dirt. All that was missing was the account of how dust was pressed into the shape of Man, how the infant was fashioned as an independent object, as a space and form apart from nature.

Reconstructing causes of death

In 1839 in Britain, deaths were divided into those of internal or external cause, a distinction that roughly followed the old coroner's separation of natural deaths from those caused by human agency. Thus, external causes were those, such as violence, that came from outside the body while internal causes were those deaths that in an earlier epoch had been 'natural' but were now classified as pathological. This fundamental reconstruction in the analysis of the cause of death, however, could not hope to be achieved in the first years of registration. There were many descriptions that fell outside the classification system and in these cases exhortations were made and letters were written to encourage a more precise terminology from the certifying doctor. From the very start of death registration causes such as 'decline', 'long illness' or 'cold' were judged inadequate but as the Registrar-General's own classification changed, so formerly acceptable designations became imprecise labels. In addition, another group of causes of death were, by the Registrar-General's own admission, 'of uncertain or variable seat'. The causes 'of uncertain seat', such as atrophy and old age, were labels that belonged to the older forms of analysis, an analysis that implied the presence of the natural processes of decay that settled a natural body back into a natural world. Nevertheless, the Registrar-General recognized that it would take time to move all deaths into the new domain of pathological causes. The label of death from old age, for example, he suggested, might be preserved until 'considerable progress is made in the diagnosis of the diseases of old people' (Registrar-General 1839: 107).

When the earliest analyses of deaths by various ages were carried out, it was found that large numbers of deaths under the age of one were classified as 'of uncertain seat'. Most of these deaths were said to be from atrophy, debility, malformation and sudden causes. Of these various imprecise causes of infant death, only congenital malformation seemed to be exclusive to children. Atrophy, debility and sudden death were equally found as causes of death – together with 'old age' – amongst the elderly. The Registrar-General recognized this correspondence between infancy and old age in 1855 when he removed these deaths from 'uncertain seat' and created a new sub-classification of 'diseases of growth, nutrition and decay'.

Within the new category of diseases of growth, nutrition and decay there were four causes of death. Congenital malformations embraced those deaths in which the evolution of the embryo had been arrested at some earlier form; premature births and debility pointed to a birth before its proper moment or to the lack of 'vitality' in a new-born baby; atrophy referred to a wasting and loss of substance without any discernable disease and was found in both infancy and old age; finally old age itself, the result of decay, made up the fourth category in the framework.

At this point, the infant was clearly an object of the natural order. In 1869, when the Registrar-General described developmental diseases it was in terms of slipping back to nature: 'The child is prematurely born, is ill-formed, feeble; the mother perishes in giving birth to her children; or the body and its elements fade away' (Registrar-General 1869: 222). It was still the great irreversible laws of growth and decay that governed infant mortality – and therefore dominated infancy. In congenital malformations the embryo, whose evolution was under the control of nature, stopped growing. It was not that the growth was pathological – as it would be in later years – but simply that it stopped too early. Prematurity represented a similar failure of natural timing when birth took place before its proper moment, and debility was a lack of that life force that made existence possible. Atrophy also belonged, like old age, to the natural decay of the body and was found both in the elderly when the life force was on the wane and also in the infant when the decay of ageing occurred earlier than usual.

The failure to provide a repertoire of pathological labels for infant deaths during the nineteenth century did not therefore reflect on medical ignorance or uncertainty so much as on the 'natural' status of the child. The infant was a biological object, a part of the natural world and accordingly subject to the laws of natural growth and regression that were held to characterize that domain ('the body wastes and the forces fail without any apparent disease') rather than those of pathological life. Yet if the origin of Man was to be located in infancy then the separation from nature needed clarification. In the same way that the death of Man demanded a rapid and definite return to nature, so at the moment of birth the infant had to be fully extricated from nature. The solution was to use the same mechanism that had already bestowed life on the adult, namely a pathological death. Atrophy, debility and prematurity, which implied an infant body hardly differentiated from nature, had to be replaced by pathological causes.

In the new mortality classification adopted by the Registrar-General for 1921, atrophy finally disappeared. With that gesture, the influence of natural forces on infant mortality was finally removed from the classification of infant deaths. Debility, it is true, remained from the old order but it was now qualified. Whereas previously it had been an inherent characteristic of the infant's life, its cause was now given a 'congenital' label. Such a change had little meaning in terms of understanding the underlying cause or basis of the problem but it signified incorporation of the 'disorder' into a pathological framework. Then congenital debility 'decreased rapidly as the certified cause of death' (Registrar-General 1949: 57) remaining in use, if unsatisfactorily, until 1948 when it disappeared in the sixth Revision of the International Classification of Disease. Deaths that had been caused by congenital debility were now caused by 'immaturity', 'nutritional maladjustment' and 'ill-defined': with that correction a nineteenth century relic was finally interred.

The decline and eclipse of infant deaths certified as being from atrophy and decay took place in the fifty years between the 1880s and the 1930s. It was also during this same period, from 1877 when the infant mortality rate was first introduced, to 1927 when still-births became registrable, that the 'infant' both achieved and consolidated its position as an analytically separate entity. Both of these revisions to the understanding of infant deaths served to reinforce the discrete and autonomous space of infancy, at last freed from the chains of natural forces. The final emancipation of the infant in the inter-war years was also marked by the emergence of a distinct area of medicine devoted specifically to children (in the form of paediatrics) as well as other institutional and professional configurations – psychological, educational, nutritional, hygienic – deployed around the newly fabricated autonomous body of the infant-Man.

The body of Man now had a beginning, a point of separation from nature, and, from that point of origin onwards, a pathological cause of death, commensurate with its non-natural status. The final twist of the kaleidoscope of birth, death and life was to analyse the patterning of infant deaths to affirm yet again Man's autonomous origins.

Analytic dimensions

In his Annual Reports, the Registrar-General described both the numbers of infant deaths and the proportions dying from different causes. In addition, he analysed the deaths further by sub-dividing them by other variables to establish patterns of death. For example, deaths for male and female infants were reported separately thereby illustrating the importance of the sex of the infant on its chances of survival. Similarly, deaths were analysed by urban/rural settings and by season. These three analytic dimensions, however, were fundamentally revised in the opening decade of the twentieth century.

Until the early twentieth century, sex differences in infant death rates were viewed as a biological phenomenon. In 1909, for example, the Registrar-General suggested that a single explanation for the excess of male over female deaths 'may perhaps be found in the supposition of a lesser initial viability amongst males' (Registrar-General 1909: cxxiii): as male infants were smaller they seemed more susceptible to diseases such as diarrhoea and bronchitis. Some two years later, in 1911, however, it was pointed out that in studies of frogs and aphids better feeding produced more females; hence, the argument went, with improvement in the national diet it was to be expected that more female infants would survive. In effect, the rationale for the excess of male deaths had begun to shift from an in-built error of nature to the more social problem of nutrition.

In similar fashion, different mortality rates in urban and rural settings had been part of the investigative framework of the nineteenth century medical statistician and as such had primarily referred to the sanitary hazards of town and city living. The Registrar-General noted as late as 1905 that:

> Many diseases which are highly important from a sanitary standpoint vary, as regards both prevalence and fatality, according as the place in question is either densely or sparsely populated.
>
> (Registrar-General 1905: xliii)

Then, within a few years, a more social interpretation was introduced. In 1909, for example, when the difference in mortality between county boroughs and towns was noted, infant mortality was not so much a product of the physical environment as of the social aspects of urban living: 'care of child life is apparently still neglected in many urban areas' (Registrar-General 1909:xliv). By the 1930s, this social analysis had been extended further and was reflected in the use of population densities, over-crowding and 'urbanization' as dimensions in the explanation of infant mortality.

Analysis of death by season had linked together meteorology and disease–an old alliance that continued in some form through much of the nineteenth century. Even as late as the very early years of the twentieth century, there was still a belief in the correspondence between weather conditions and disease. For example, the Registrar-General expressed this conviction in 1905 when he declared that:

> It may eventually be found that many of the ailments incidental to humanity depend for their prevalence on variations in meteorological conditions more closely than is at present thought possible.
>
> (Registrar-General 1905: xliii)

The chief vehicle for this belief, in so far as infant mortality was concerned, were the diarrhoeal diseases that were observed to have a strong seasonal variation corresponding to changes in temperature. But in 1911, diarrhoeal diseases, from being an exemplar of how the causes of infant mortality might be revealed, were relegated to the status of mere anomalies. In that year the Registrar-General suggested it might be useful to exclude diarrhoeal deaths from the infant mortality rate as their wide fluctuations with the weather confused interpretation of real trends.

Thus, within the space of a few years in the first decade of the twentieth century some of the traditional physical parameters by which death had been interrogated during the nineteenth century were fundamentally undermined. Three separate strategies were employed: in the case of urban/rural differences, a related but more social measure of town/county was introduced, in the case of sex differences the cause was reinterpreted, and in the case of seasonally-based diarrhoeal diseases the phenomenon was simply dismissed as of no significance to the proper understanding of the problem. Each of these shifts affirmed a new space for the infant: not the physical space of nature, geography and meteorology, but the separated space of the independent body.

The definition of this autonomous corporal space depended in part on its distinctiveness from nature. In addition, the analytic framework that informed the meaning of infant deaths established an internally differentiated space with social rather than natural dimensions. The crystallization of the infant in this new space can further be illustrated by the parallel emergence of two new, decidedly social, parameters, namely legitimacy and social class. The Interdepartmental Committee on Physical Deterioration of 1904 had noted the link between illegitimacy and high infant mortality (Interdepartmental Committee Report 1904) and in 1908, the Registrar-General introduced a regular breakdown of infant mortality by those 'unfortunates' born illegitimately. Then there was social class, first used in 1913 for the year 1911 when infant deaths were analysed by the occupation of their father.

The realignment of the conceptual space of infant mortality from one traversed by sanitary axes to primarily social dimensions meant that the infant, through changes in its relationship to nature and in the form of its death, had become an essentially social object.

Newman (1906) subtitled his book on infant mortality 'a social problem' and described it as 'not without national importance'. In the late nineteenth century when improvement in infant mortality signified 'progress in sanitary reform' (Registrar-General 1887: xci) there was a fear that this represented dangerous interference in the natural order such that many infants were being saved who perhaps 'should die' (Registrar-General 1881: xiii). Even in 1907, the high mortality in the first week of life was held to be mainly due to deaths from immaturity and debility among infants that could 'hardly be regarded as viable' (Registrar-General 1907). Yet in 1913, in the context of social class differences, much of infant mortality, particularly in the lower classes, was seen as preventable. By 1918, it was apparent that 'mortality in the first fortnight is not by any means entirely due to the elimination of non-viable infants' (Registrar-General 1918: xxxi).

This reconstruction of infancy from a natural state to a social one was also accompanied by a shift in the underlying form of the explanatory framework that sustained this new vision of infancy. When a monolithic physical environment was the principal discriminatory structure, mono-causal models predominated: it was dirt, or, later, germs, that caused disease. With the interpolation of social factors, however, a more dynamic and complicated framework was established. In 1912, a mechanical system of tabulating death certificates was introduced to enable identification of the alternative ways that the cause of death might be defined. The Great War delayed this process but interest continued in the inter-war years. In 1927 a new death certificate was introduced in which the old rubric of primary and secondary causes of death was replaced with a threefold classification of an immediate cause, others (of which the immediate cause was a consequence), and any other unrelated contributory diseases. Although it was not finally accepted until 1940, the old method, by which certain arbitrary rules governed the selection of the primary cause of death whenever more than one cause was mentioned on the death certificate, was superseded by selection based on the opinion of the doctor who issued the certificate. By 1954, multiple cause analysis was an established procedure.

In summary, it took only a few years at the beginning of the twentieth century to transform the conceptual space in which infants were located. The old axes of climate and urban living that reduced a population to a collection of separate bodies gave way to social dimensions as the corporal space of infancy was finally wrenched fully free from nature. However, it took several decades to seal the transformation; it was only in 1954 that the Registrar-General finally discontinued the old practice of providing an extensive Meteorological Report and directed interested parties to the government Meteorological Office should they have need of such data.

These steps of consolidation meant that by the middle of the twentieth century Man not only had a point of origin but an explanatory framework for achieving and legitimizing the momentous shift from the natural world to the social. In this sense the 'problem' of infant mortality was not a historical constant; it did not lurk on the underside of society waiting to be discovered by an enlightened public, but was invented by an analysis that established its existence both at that moment and, through retrospection, in the past. This creative process established a beginning for Man no less significant than the ancient skull fragments of some hypothesized 'missing link'. To be sure, this creation story was not of a mythical ancestor, of an Adam and Eve, or even of a Lucy, yet it still provided an account of the very tangible origins of every Man.

The modern search for beginnings might be said to have begun with Darwin's *Origin of Species* of 1859 and *The Descent of Man* of 1871 that posed the daring hypothesis that Man's origins lay far in an evolutionary past. Yet there was this other search for beginnings – different yet more pervasive – that occurred roughly contemporaneously when the Registrar-General in 1857 first identified infant deaths as a particular problem. Could it only be coincidence that the search for the origins of Man was accompanied by the search for the birth of Man? For Darwinism the evidence lay trapped in fossils and rocks laid down millennia ago; alternatively, it is possible to place the beginning of Man, that discrete space of independent identity, in fertile texts rather than in the barren earth. Why else did Darwin's *Origin of Species,* Gray's *Anatomy* and the Registrar-General's early quest for the beginnings of life emerge within the same few years of the nineteenth century? The crucial question was indeed – as the Darwinists recognized – Man's place in nature, but the answer was not to be found through discovering some putative 'missing link' with the great apes, but through dragging the infant clear from the womb of nature. The birth of the infant had become the origin of Man.

5
Making the Body Move

In the nineteenth century, the space of 'nature' had opened up a new plane in which the body of Man crystallized as a solid, discrete and analysable object. Sanitary science, through its concern with monitoring the boundary between nature and Man, played a central role in the mechanisms that separated corporal space from non-corporal. Yet these techniques of separation served only to establish the human body as an immobile object. This was indeed an anatomical image of Man, reproduced in countless descriptions in anatomical atlases and texts, a constant structure fixed in the 'anatomical position' – body upright, arms by side, palms facing forward – so that its topographical coordinates could be precisely mapped. The emergence of a more dynamic body in which corporal boundaries became more flexible and permeable had to wait until the twentieth century.

Sanitary science had defined a three-dimensional body largely through techniques that forced a fissure across a two-dimensional field. In essence, the body boundary was little more than a line that allowed spaces to be characterized as being on one side or the other. In the twentieth century, however, the body began to become truly a three-dimensional space through the adoption of a new set of techniques that came to direct and order a changing shape. Exploration of the body's spatial dimensions would reveal its latent power to move – slowly and methodically at first, but soon with speed and freedom. It was not that motion was absent from nature, only that the newly fabricated body could not rely on natural forces to impel it forward. Man had to be animated through a new set of movements; he had to be given a new repertoire of bodily gestures that were far removed from the impersonal forces of nature.

The army had discovered that getting men to move rhythmically in unison enabled raw recruits to be transformed into disciplined soldiers. The beat of the drum integrated marching soldiers into a military corps; the weapon drill that coordinated bodies and rifles forged a unified fighting machine. A 'technological breakthrough' occurred in 1889, however, when a Captain Fox introduced a major revision to the traditional pattern of drill in the British Army: drill without weapons. Weapons drill, it was argued, over-developed the arms and chest without compensating balance from the legs and trunk. The new intention was to develop the soldier as a whole by using the body as the object of drill not the body as an adjunct of the weapon.

The development of drill without weapons – or physical training, as it was to become known – established techniques of body management that could easily be transferred from military to civilian life. And what better place to start than the growing body of the child. The army formalized its new approach to drill with the publication of a *Manual of Physical Training* in 1908; the Board of Education published a *Syllabus of physical exercises for public elementary schools* a year later.

> It is especially during the period of growth, when body, mind and character are immature and plastic, that the beneficial influence of physical training is most marked and enduring; and the highest and best results of education cannot be attained until it is realized that mental culture alone is insufficient; and that physical education is necessary to the development not only of the body but also of the brain and character.
>
> (Board of Education 1909: 1)

The stage was set for movement to begin to flow through the inanimate three-dimensional corporal space of the body.

Posture

One of the most important components of physical training to be placed at the heart of the elementary school was an emphasis on posture. Posture was not movement *per se,* but it was an essential precursor. Posture addressed both internal and external aspects of body management; at once it focused attention on the internal alignment between body parts and it also enabled the three-dimensional shape of the child's body to be turned towards an outside world. For example, the *New Model Course of Physical Exercise* of 1904 took the old military position of standing to attention and applied it to the body of the child.

> The body and head must be held erect chin slightly drawn in, chest expanded, shoulders square to the front and slightly drawn back, and eyes looking straight forward. The arms must hang easily, elbows to the rear, fingers and thumbs straight close to one another, and touching the thighs; knees well braced back, heels closed, and toes turned out so as to form an angle of about 90 degrees; weight of body on the fore part of the feet.
>
> (Interdepartmental Committee Report 1904: 17)

Standing to attention used attention twice over: it involved, literally, an attention to the positioning of all body parts so that all were correctly placed in relation to each other and the world outside. Attention also reflected the body's preparedness, its readiness to wait and listen for commands that would fire corporal space into motion. Standing to attention was therefore the ideal posture, the point of highest perfection for body positioning and alertness to subsequent movement. It was, of course, a position that transcended nature: an over-caricatured version of natural alertness that replaced anticipatory tension with a taut formality. Indeed, the position of attention was pre-eminently 'unnatural' in the control it exerted simultaneously over every body part and muscle. How better to wrest the body from natural forces? Alas, this posture could only be held for brief periods and the rest of the time a child's body relaxed and departed to varying degrees from this great ideal. Nevertheless, while standing to attention might be an infrequent activity, posture became a constant outward expression of the state of inward control of the child's body.

If a child slouched forward then this weakened the neck muscles, rounded the shoulders and caused injury to the eyesight; if a child slouched backwards, it produced a curved spine. Moreover, both forward and backward slouching constricted the chest and impeded breathing. Folded arms also affected movement of the chest as well as encouraging stooping, but arms held behind the back could be equally harmful in producing unnatural stretching of muscles of the shoulder, chest and back. The best place for the arms, it was advised, was down the side of the body resting on the hips. In effect, an essential precursor to movement was body deportment, each position deconstructed, critically analyzed, and an ideal declared. When Porter (1906) argued for the importance of posture because the child's body could so easily 'depart from the proper attitude', the idea of a proper attitude did not derive from empirical observation still less from the notion of a body as a product of nature, but from a properly analyzed and positioned physical presence. Is it surprising that the concept of 'attitude' that 30 years later was to become a cornerstone of the expanding discipline of psychology should have been forged in these early explorations of body positioning?

In the early years of the twentieth century, the focus was on the educational and training implications of posture, but later more explicit links with health were made. Taylor placed posture first in his list of the five aspects of physical exercise that related to health (attention to posture, mobility exercises, muscle function, breathing exercises and skin friction – he suggested rubbing the skin with a goat hair glove). He noted that some American universities went so far as to take frontal and lateral sillouettographs to identify 'postural blemishes' so that students might be 'shocked or stimulated' to take corrective action (Taylor 1934). Even minor postural 'defects' such as flat feet became more the province of medicine than education (Phillips 1934).

If attention to the body's posture can be seen as an important precursor to movement itself so can the preliminary checking of locomotive potential. Movement would throw demands on muscles and breathing so these had to be placed in a state of readiness. (Such readiness for action was no better illustrated than in the Scouting motto: 'Be prepared'.) Atkins praised the Model Course for introducing exercises that would teach 'the extremely important art of breathing' (Atkins 1904). Newton advised that a 'thorough and painstaking examination of the whole body and of each particular function of every pupil should be regularly made, and a complete chart should be filled out ... ' (Newton 1907: 666). Milligan (1918) even suggested a national Central Anthropometric Bureau where such statistics could be registered.

Exercise

Then the body could be brought into action. Not hurriedly, but building up to its full potential. Brunton (1915) thought that the best exercise was one 'which puts in action every muscle of the body, but does not put any one into action for too great a length of time at once or in too violent a manner' (p. 3):

> Slow exercises requiring a certain tension of the muscles, such as posturing, dumbells, Indian clubs and the use of elastic cords tend to increase the strength of the muscles. All kinds of play with throwing and catching balls increase the coordination by which eye, body and limbs work together, while running tends to develop the lungs and heart. The directing power of the brain is increased by drill, which teaches the child to go through movements at the word of command. The best system of physical education is that which will meet all the necessities of the various organs of the body.
>
> (Brunton 1915: 3)

With this regime, the muscles could be increased in strength and endurance, the heart made stronger, the circulation more active, with additional indirect effects on the power of digestion and on the mind. The result was a 'sensation of well-being, bodily and mental, (that) enables the individual to resist all the previous destructive agencies, microbic and climatic, which tend to produce disease' (Brunton 1915: 3).

Movement was therefore not a casual thing, even less an attribute that could be left at the mercy of 'natural' forces. It needed to be directed and managed, subsumed even, under some explicit inner and/or outer controls. Movement required order and guidance. Certainly, children were born with muscles but these needed bringing under the body's wilful control:

> In all exercises muscles are being controlled by the will. The oftener the centres are called upon to exercise their controlling influence the better they can do it and the more perfect becomes the control.
>
> (Porter 1906: 40)

With a prepared posture together with attention to efficient breathing (with shoulders squared and mouth closed), the child was placed in a state of readiness for movement. Then the basics followed: rhythmical swinging arms, correct positioning of the head, the management of body bearing, the coordination of legs. Drill routines – 'marching, counter-marching, diagonal marching, changing ranks and so on' (Atkins 1904: 3) – embodied many of these exact movements and postures, increasing the directing power of the brain so enabling children to go through movements at the word of command. It was a question of getting the right balance between prescription and self-direction:

> A good neutral system of physical training ... with enough control to produce discipline, and enough spontaneity to encourage originality, is to be regarded as an integral part of a sound elementary education.
>
> (Atkins 1904: 3)

The disciplining of bodily movement in school was widened from the formality and rigidity of drill to the more generalized techniques of physical training and physical education both of which were underpinned by carefully graduated and scientifically calculated movements (Atkins 1904). The individual body could be trained, particularly through repetitive actions - so inculcating 'habits' - to move towards a goal of corporal fitness and efficiency. Training took control over the general physical development of the child ensuring that full physical capacity was achieved (Newton 1907).

The syllabus for the Model Course of 1904 described the value of exercise. First, exercise had a number of physical effects. It improved general nutrition and it had a 'corrective effect' in terms of the remedy or adjustment of any 'obviously defective or incorrect attitude or action of the body, or any of its parts' (Interdepartmental Committee Report 1904: 3). Further, it possessed a developmental effect whereby the body as a whole attained the highest possible degree of all-round fitness particularly through the accompanying specialization of brain cells concerned with accurate and coordinated movement. Second, exercise had an important educational effect. It had a 'strong mental and moral' impact and played an important part in the development of character.

> Rightly taught, Physical Exercises should serve as a healthy outlet for the emotions, while the natural power of expressing thought, feelings and ideas by means of bodily movements is encouraged and brought out. ... This appeal to the aesthetic sense is very great, and extremely important, for in learning to appreciate physical beauty in form and motion, the perception of all beautiful things is insensibly developed and the child gradually learns to seek beauty and proportion not only in his external surroundings, but also in the lives and character of those he meets.
>
> (Interdepartmental Committee Report 1904: 5–6)

A significant weakness of early forms of physical training was their reliance on external systems that achieved their effect through habit forming ('the child unconsciously acquires habits of discipline and order, and learns to respond cheerfully and promptly to the word of command' (Board of Education 1909: 1)). The more desirable alternative was to instil the word of command into the child itself so it could, in effect, function as its own drillmaster. This emphasized the role of education in body management so the child could also equally well – and undoubtedly more efficiently and permanently – train his or her own body. Thus, a virtuous circle linking body, mind and exercise was established in the early decades of the twentieth century. The mind could be trained to manage the movement of the body while movement, in its turn, had positive effects on the mind. Exercises could inculcate habits while the discipline of movement could focus the mind on the task in hand, eliminating stray thoughts – perhaps of a sexual nature – and build up 'character' (Hussey 1928: 578). There was apparently 'a very close relationship between intelligence and success in athletics' (Ruble 1928: 216), and mental problems might be expressed in poor movement coordination just as true character could be read from the sport's field, the gymnasium, the athletics ground and the dance school:

> For every physical expression there is a mental equivalent. ... By watching the action of the individuals in (the class) a shrewd insight into their state of mind, temperament and character can be obtained.
>
> (Campbell 1940: 351)

Physical training in all its varied forms became emblematic of the twentieth century concern to take the still, inanimate body of the nineteenth century and give it movement. Tested in the classroom, the link between body, mind, and movement spread in the inter-war years to the world outside as muscles were stretched and bodies moved in disciplined unison. Posture, fitness and efficiency were the initial objectives but with the gradual extension of the techniques through which the body was manipulated the goal widened towards a regime of total health. From the growing emphasis on physical training and education in schools to the value of exercise in convalescence (Barron 1916), from the increasing popularity of the revived Olympic Games to the growth of hiking, camping, municipal tennis courts, baseball diamonds and swimming baths, weekend football, athletics, and cycling for everyone, from dance for girls so they might develop grace (Chisholm 1925) to massed displays of physical drill (particularly in continental Europe), movement and exercise rippled throughout the formerly rigid corporal space of the nineteenth century. Fed back into universal military training, the disciplined soldier could be replaced by the trained body:

> Defective vision could be corrected; men with weak feet and backs could be relieved. The mental defectives could be classified; the young man would be given an object lesson in discipline and be taught respect for authority–the good that could be accomplished is unlimited.
>
> (Ireland 1920: 499)

The defective body, riddled with 'physical deterioration', unable to fight in wars or work in industry, could appear as a new object of social and health policies. From a strategy of eugenics to a system of corporal training, the body of everyman and everywoman could aspire to perfection.

> Self-reliant, self-disciplined intelligent individual to replace the dragooned type of person who may be well disciplined but is mainly guided by an adolescent mind trained to conditioned reflexes.
>
> (Cove Smith 1940: 161)

The transformation of those passive nineteenth-century bodies was remarkable. Frankenstein's monster was learning to move. The stretching and flexing of limbs represented an escape from the strict confines of the sanitized body of the previous century. This body was finally free of nature and could extend its corporal space by using movement to map out, both physically and metaphorically, its unbounded domain. It was a confident body that could now begin to re-engage with nature, not as a small cog in the vast machinery of the natural order but as an autonomous object. In a novel turn on nineteenth-century miasmatic theory (which held that impure air was a cause of illness), nature – no longer a source of danger – could be allowed to invigorate the body with its fresh air. Bodies could sleep in the open or under canvas and walks or 'hikes' could be conducted across a terrain that was as purely natural as could be found. City life meant 'biological death' (Abrahams 1930). Just as a body free from dirt had separated the figure of Man from nature in the nineteenth century so a body that was fit and able to re-engage with nature defined an essential element of Man in the inter-war years of the twentieth century. What then could be more logical than a eugenics policy to remove those whose fitness compromised their human identity? What more rational than a new sanitary regime of purification that addressed bodies rather than dirt?

'Back to nature' was a slogan that was only possible as a consequence of the enormous journey away from nature that had been accomplished in the previous century. It was not the realization of a Darwinian dream, of a body adapted to its environment for that would require compromise and subservience to a natural order from which the caesural break had finally occurred. Indeed, adaptation rather than engagement was an often unseen danger:

> 'Recently we have come to appreciate how deceptive a child's outward appearance may be; if malnourished or insufficiently provided with fresh air, sleep and exercise, its body will attempt to compensate for this lack and it is precisely from these compromises that the later troubles spring.
>
> (Cove Smith 1940: 159)

Anatomical space, so critically fashioned in the nineteenth century, had moved even further away both from the world of the natural and from its former stillness towards the active autonomous subject that was to seize the body in the latter part of the twentieth century.

In that sense, the precise corporal space of the nineteenth century was beginning to disappear at the moment of its perfection – just like H.G.Well's invisible man of 1897 – as the fixed carapace of the body began to flex. At the same time, the cutting edge of sanitary science, which had constructed the early icon of Man, was replaced by a new regime of hygiene that was concerned with dangers arising from other places and other spaces.

6
Creating a Social Identity

By the early years of the twentieth century, the struggle to emancipate the body from nature had largely been won. The contours of the body defined an autonomous object located within, yet separated from, a natural landscape. This positioning implied that danger was located along the interface between body and nature, a threat constantly reaffirmed by a sanitary science that guarded and maintained the body/nature divide. In the early twentieth century, however, the boundary of the body was realigned, and a new space of danger was revealed to the watchful eye of medicine.

Animation of the body meant that its spatial limits needed to be redrawn. The elements of drill had formalized the potential space of one body in its relationship to another, but with increasing emphasis on free movement, body space could no longer correspond to the fixed limits of the anatomical position nor could coordinated movement manage it. Limbs and bodies needed to move freely; yet this very freedom posed dangers of mixing with the 'space of action' of other nearby bodies. Indeed, the problem can be expressed as how best to manage inter-personal space. The danger was not from an unsanitized nature passing by means of dirt into and out of the body's volume but from a body intruding into the widening space of another body. In the opening decades of the twentieth century, the gaze of public health therefore began to focus on a new target, namely the space between bodies. 'The old public health was concerned with the environment;' noted Hill in 1916, 'the new is concerned with the individual' (Hill 1916: 8).

Initially, the space between bodies was a physical gap across which dangerous substances could spread. Tuberculosis, for example, had been a disease of physical environment, of dirt, poor housing and insanitary conditions, but in the early twentieth century, it became a disease of social contact, of breathing, coughing, and proximities. Regular and voluble clearing of nasal and respiratory passages, a routine part of everyday life, was designated as unclean; increasingly spitting became socially proscribed and spittoons were removed from public places. Venereal disease, for long a mark of immorality, became the archetypal manifestation of the dangers of interpersonal contact. Commissions, clinics, and public campaigns highlighted the incipient dangers of a disease that spread from body to body throughout the population and a network of medical personnel was established to retrace its secret passage from one corporal space to another. With these new strategies, sanitary science began to mutate into a hygiene of inter-personal space as public health fixed on a new constellation of dangers.

Inter-personal hygiene

It is perhaps ironic that twentieth century society claimed to pay such scrupulous attention to safeguards against laboratory and clinical experimentation on children when the child was so often the principal target of trials of a more fundamental nature. The school might have been established as a place for learning, but it also functioned as a laboratory in which the body of the child could be subjected to analysis, experimentation and transformation. Where better to apply and test out the new principles of inter-personal hygiene?:

> The school child, easily seen, easily examined, easily described has enabled us to crystallise the conception of personal hygiene and to test the possibilities of remedial measures.
>
> (Mackenzie 1906: 510)

School hygiene had first emerged in the latter half of the nineteenth century as a part of the wider regime of sanitary science. It became a specific, if relatively minor, interest of the school medical inspector who sought out dirt and applied the principles of sanitary science:

'water, drainage, lavatories, cloakrooms, systems of heating, systems of ventilation, fire, opening and closing windows' were all part of the inspection ritual (Mackenzie 1906: 6). In effect, schools were yet another place for the application of the sanitary rule that required dirt to be kept separate from bodies. At the turn of the twentieth century, however, a new hygienic concern appeared: one child could be a source of danger to another through contagious disease. The school thereby became a place of potentially dangerous contacts and exchange. The school was sited in the midst of the community mixing children from their separate domestic spaces so that disease in one home was quickly transferred to another:

> There is a mass of evidence showing conclusively that the schools are a principal means of disseminating disease throughout the community.
>
> (Gulick and Ayre 1908: 50)

Following the advice of the medical inspector, certain diseases were labelled as 'excludable', though there was considerably variation in this; some schools excluded the common cold, for example, while others did not. On identification, the contagious child was then prevented from attending school; a formal exclusion notice was sent to the parent and records completed for the school and the health authority and for the medical inspector himself.

The system of exclusion that spread through schools around the end of the nineteenth century resembled the old system of quarantine in which separation had maintained order and purity. Whereas quarantine acted to keep places separate, school exclusion kept children's bodies separate from potentially dangerous consociations. Sanitary science resonated in the new system of exclusion in that the main threat still came from dirt. The child:

> is collecting and redistributing dirt and dust all day long ... the restless, growing skin-shedding, mucous shedding, dust distributing, spitting, coughing, and shouting demon that the school boy, at his worst and best, always is.
>
> (Mackenzie 1906: 5)

But this was a new expression of dirt. Under sanitary science, dirt originated in a corruption of the natural environment; sanitary science, for example, had advised that schools should be properly sited in an open position on clean dry soil, not on 'made soil' constituted by refuse and road sweepings. In contrast, under the new regime of inter-personal hygiene, dirt came directly from the body itself, from the other child.

By 1905, the new focus on the child's body boundaries in relation to other children was beginning to overtake the traditional sanitary concerns of school inspection. 'We are entering an era of Personal Hygiene' observed Mackenzie (1906: 50). Within a few years, the change was complete.

> Personal hygiene has been taught to children during the last 2 or 3 years in a manner and with a force never approached before. The value of personal cleanliness, of the major care of the teeth, and of exercise; the nature of infectious disease, and the deleterious effects of alcohol, coffee, tea, and tobacco have been taught vigorously.
>
> (Cornell 1912: 1)

The innovative feature of inter-personal hygiene was the attention paid to the spaces between people. Most obviously, this was a physical space across which contagious organisms passed; it was a space of movement, of transmission, of exchange. For example, pupils in Providence, Rhode Island, were instructed on the dangers that lurked on body surfaces:

> Do not put the fingers into the mouth. Do not pick the nose or wipe the nose on the hand or sleeve. Do not wet the finger in the mouth when turning the leaves of books. Do not put pencils into the mouth or ... money ... (or) pins ... (or) anything ... except food and drink. ... Apple cores, candy, chewing gum, half eaten food, whistles or bean blowers were not to be swapped; pupils (were commanded) never to cough or sneeze in a person's face and to keep hands and face clean.
>
> (Gulick and Ayres 1908: 57)

In 1925 Mauss, the social anthropologist, claimed that the affinity between bodies could be symbolically expressed in the social process of gift exchange. Thus, while the new inter-personal hygiene policed the transmission of physical matter between bodies, it also, inevitably, addressed the underlying space of relationships. The idea of exchange recognized that inter-corporal space often reflected a relationship between two bodies, of friendship, or of relatedness, or of intimacy. The space might therefore be a physical one but it was also an inter-actional one, emanating from one person to embrace another. Therefore, as a reinvigorated public health annexed this new empire it extended further the limits of individuality: to the anatomical identity of sanitary science was added the relational characteristics invoked by inter-personal hygiene. How better to describe this emerging space of relationships than 'the social'? The social as a noun, as an analysable object, crystallized in those early years of the twentieth century when numerous new bodies of knowledge formed around the ever-unfolding space between bodies. The founding professor of sociology, Emile Durkheim, might have been first to define and map out this autonomous realm at the beginning of the twentieth century, but public health was also quick to seize upon the hygiene of this diffusing inter-personal space and designate it 'social medicine'.

A medicine of the social

In his overview of the emergence of social medicine, Sand identified its first appearance in the early years of the twentieth century with the advent of socio-clinical medicine, industrial medicine, insurance medicine, preventive medicine, and social hygiene (Sand 1952). There was a professorial chair in the new discipline in Germany as early as 1902 and by mid-century, it had spread throughout Europe. In 1942, the Royal College of Physicians of London set up a Social and Preventive Medicine Committee to prepare for the development of a subject that offered 'a relatively novel point of view'. The Report of the Interdepartmental Committee on Medical Schools, chaired by Sir William Goodenough and published in 1944, acknowledged that Social Medicine:

> signifies a particular conception of Medicine; a conception that regards the promotion of health as a primary duty of the doctor, that pays heed to man's social environment and heredity as they

> affect health, and that recognizes that personal problems of health and sickness may have communal as well as individual aspects.
> (Interdepartmental Committee on Medical Schools 1944: 167)

Specialization in medicine had been predicated on the medical task itself, on sub-dividing the space or 'volume' of the body. Social medicine, however, introduced a new organizing principle for the medical division of labour; social medicine was different precisely because it was not directly related to the analysis of corporal space but to a fertile space outside the body that would, in time, provide the basis for a fundamental reorganization of the basic notions of health and illness.

In the nineteenth century, public health in the form of sanitary science had focused on the corporal space of the patient located in a wider space of 'nature'. The latter was a residual space, more important for what it was not than what it was. In the late twentieth century, this contextual space would be formalized and given a significant identity, but in the first decades of the century, it was the emergence of a new inter-personal space, strictly neither corporal nor non-corporal, that caught medical attention. Social medicine – or preventive medicine as it was known in the US – did not engage with an isolated physical body, or indeed a simple collection of those same bodies; instead, it recognized the interactions of those bodies as the locus of illness and as the space for intervention. The problem was no longer the line that separated corporal and non-corporal space but the space – both physical and social – between one body and another body. Just as sanitary science had ushered in a new hygienic regime that separated one object (the body) from another (non-body) so too the twentieth century witnessed a reconfiguration in the hygiene strategies that surrounded and empowered the extended space of personal identity.

The emergence of a new social identity involved a focus on both the space between bodies and movement across the increasingly multi-dimensional void. Both bacteriological theories of contagion and Freudian theories of transference addressed this new inter-personal target. The new project was to shake up the rigidities of the old fixed and delimited corporal identity and fracture its former boundaries. The effects of this new focus were felt throughout the social body. For example, the new social medicine embraced activities as varied as:

> aptitude tests in the recruiting of labour, the fighting forces and auxiliary services, civil as well as military; systematic training for the work proposed; rehabilitation of the disabled; rational distribution of foodstuffs; extra health protection, particularly for workers, expectant mothers and children; organization of leisure activities; and development and improvement of social services.
>
> (Sand 1952: 3)

Sanitary science had focused on the anonymous boundary that separated the mass of bodies in the population from their external environment. Personal hygiene began to recognize countless individualities in that formerly undifferentiated anatomical accretion, each composed of different constitutions and habits:

> Too much stress cannot be laid upon the fact that it is the constitution, the nature of inherent tissue, that controls or modifies the inception of many of the ills of life. ... The principles of personal hygiene may be more readily taught to and inculcated in the young, but with much greater difficulty can we affect the mature or aged; for we are all creatures of many habits, and in the mature adult the impress of these may resist to the utmost any and all endeavours to modify or remove them.
>
> (Egbert 1903: 271)

Instead of a mass of uniform and inert anatomical figures, the new analysis of inter-personal space identified individual differences. Children could be ranked by intelligence and by gymnastic skill; adults could be differentiated by personality and physical prowess. General hygienic rules and regulations were:

> well recognized and of the utmost general value, (they) cannot, however, be stated in very specific terms when applied to individual conditions because the individual idiosyncrasies of different persons vary to such a great extent.
>
> (Bergey 1904: 260)

The problem for public health was that the identifiable hygienic line of separation that characterized sanitary science had become a multi-dimensional space that diffused into individuality. This social space was part of an extended body yet was also separate from it, and shared. It was as if the discrete corporal spaces of the nineteenth century were dissolving their physical boundaries and beginning to coalesce with neighbouring bodies. The social could be held neither within nor without the body by hygienic rules of exclusion; the challenge for social medicine was therefore how best to maintain hygiene across the mingling space defined by these permeable body boundaries. A person could not be kept in perpetual solitary confinement; they would always be surrounded by the dangerous space of contagion, by the threats, physical, psychological and social, that lurked in every relationship. The solution was to redirect monitoring to those spaces in which potentially dangerous social mixings could occur, a strategy of constant hygiene at work, in the home, and during leisure activities. With this new focus, the old hygienic concerns were dissipated: public health interest in climate, soil and building would hardly survive the end of the nineteenth century and problems of monitoring clean food, air and water and disposing of effluent would be relegated, as the twentieth century progressed, to a subordinate, technical and routinized administrative function.

In his advice of 1919 on public education in hygiene, Newman had commended the practice of hygiene, teaching of mothercraft, physical education and open-air education (Newman 1919). These elements defined a new space in which medicine could operate. It also provided a new target for and effect of those practices. Interpersonal hygiene and social medicine addressed the new dangers that lurked between bodies, which had their origins in the body of another. Mothercraft brought the individual's first fundamental relationship within the ambit of physical hygiene just as surely as Freud was bringing it into the realm of mental hygiene. In physical education, the ordinary bodies of everyone were moved and stretched together, mapping out the expanding suzerainty of the social in the open spaces that had once been the exclusive preserve of nature.

The surging space of the social transformed Man. No longer coterminous with three-dimensional corporal space, identity decomposed into 'individual idiosyncrasies of different persons'. Moreover, the application of inter-personal hygiene and social medicine to the different domains of body interactions mapped out the analytic features of social life itself, public and private, work and home, in which identity was crystallized. This new identity was without distinct boundaries, precarious and constantly re-negotiated. It was fragmented across different social milieu, constituted and reconstituted by the affinities with other identities, a different role for each segmented space of social life.

This new identity for Man only existed in relationship to the Other, the defining yet anonymous figure that lay on the other side of interpersonal space. Indeed, the first half of the twentieth century was marked by the search for and analysis of this shadowy twin who enabled the realization of Self. From Piaget's children learning to separate themselves from others to Laing's tortured divided selves, the Other moved from being, by definition, the non-self, to a central part of identity. Between contagious theories at the beginning of the century to the formalization of the 'problem' of the Other in texts such as Wilson's *The Outsider* in 1956 and Becker's *Outsiders* of 1963 or his *The Other Side* of 1964, the self flowed out to fill the new space that diffused out of the body's discrete anatomical volume.

7
Invoking Subjectivity

During the first half of the twentieth century, the inert body that had been constructed in the previous century became animated. It postured; it moved. In so doing, it began to establish its uniqueness. From beauty contests to sporting championships and from nudism to bodybuilding, the shape of bodies became both a private obsession and a public spectacle; but these body displays were only the outward expressions of deeper differences in identity. Body boundaries no longer defined the limits of individuality as the social space of interaction burgeoned outwards; it was in this new space that subjective identity began to coalesce.

In the nineteenth century, there were a number of methods – from the cross-sectional anatomical drawing to the boundary-maintaining rituals of sanitary science – through which the form of the physical body could be mapped. Similarly, in the early twentieth century the existence and dimensions of social space could be inferred from the movements and threats that crossed this emerging multi-dimensional volume. So how best to look into the psychological space of identity to witness change? Ideally, it requires a probe to enter the hidden space of the nascent mind and return a reading on its changing state. Therefore, just as the clinical examination had been developed to reveal the pathological secrets of the inside of the body, it is the evolving techniques for interrogating the patient's mind that provides a point of access to the developing locus of personhood.

The clinical examination

The medical revolution of the late eighteenth and early nineteenth centuries had brought about a radically altered perception of illness. Instead of pursuing a shifting collection of symptoms, medicine began to localize illness to a pathological lesion, a specific abnormality of structure (or, later, function) situated somewhere in corporal space. This reduction of illness to a pathological lesion was the great conceptual innovation that transformed medical understanding and clinical practice. Thereafter the medical task was directed to the identification (followed by treatment, if possible) of this new materialization of the formerly intangible symptoms of illness.

Identification or diagnosis of the lesion was based on the effects it had on the body. The patient experienced the negative impact of the lesion as symptoms and was able to report these to the doctor – a pain, a cough, shortness of breath, a fever, etc. The real innovation of the new pathological medicine, however, was the development of techniques that allowed the clinician to bypass the patient's self-report and attempt to identify the exact nature of the lesion from the tell-tale indicators or signs it left within the body. There were four fundamental skills of the clinical repertoire: inspection, palpation, percussion and auscultation. Inspection involved scanning the body with the eye: Were there signs of consciousness? How easy was the respiration? Were there any visible marks or lumps? Palpation was carried out by laying on hands to feel for the presence of any abnormality of structure. Percussion required a tapping on the body to be able to distinguish the different densities of pathological lesion and surrounding structures. Finally, auscultation allowed the ears to be used in conjunction with the newly invented stethoscope to listen to internal processes, such as respiration or the beating of the heart.

The ability to 'read' the bodily signs of the lesion of which the patient might be unaware – perhaps a mass, or a point of tenderness, or a skin colour, or an irregular pulse – was coupled with knowledge of the different manifestations of each disease type. This allowed the significance of a sign to be assessed in terms of its likelihood of indicating a particular pathological lesion. In addition, the symptom offered collaborative support in inferring the exact nature of the underlying disease.

Of the two clues to the nature of the pathology, the symptom and the sign, the latter was the most highly prized. It was as if the sign allowed the lesion to speak directly to the clinician's senses whereas the symptom was more indirectly passed via the patient and was accordingly judged less reliable. The ability to elicit the sign was the defining mark of clinical acumen that enabled one clinician to be ranked higher than another. In contrast, enquiring after the symptom seemed hardly a skilled activity. Indeed, the clinical teaching manuals that were published in the nineteenth and early twentieth centuries reflected the dominance of signs in medical diagnosis, barely mentioning the process of obtaining reports of symptoms (the medical 'history') from the patient. For example, Stevens' *Medical Diagnosis* of 1910 offered only three pages out of 1500 to the 'interrogation of the patient' (Stevens 1910). Cabot's *Physical Diagnosis*, from its first edition in 1905 until its 12th in 1938, ignored even a token statement on interrogating the patient, concentrating entirely on the physical examination (Cabot 1905). Emerson's *Physical Diagnosis* of 1928 could only manage a one-page outline of advice on the lay-out of the consultation room as a preliminary to the details of the examination (Emerson 1928).

In the various texts that did offer advice on interrogating the patient the format was almost identical. It was advised that the patient's age, sex, occupation, address and marital status should be noted before asking about the main complaint and its duration. This was to be followed by questions on the patient's previous medical history, family medical history and the so-called 'personal history' that tended to cover health hazards of the occupation, past residence abroad and 'habits' such as consumption of tea, alcohol and tobacco.

These descriptions of how to invite reports of symptoms indicated the contemporary way in which medicine perceived the identity of the patient. Eliciting the patient's story was a secondary activity; clinical skill was mainly directed at the much more valuable signs of disease that lay in the patient's docile body awaiting the inferential techniques of clinical perception. This application of the repertoire of the physical examination helped identify and realize a corporal space, a three-dimensional volume of organs, tissues and cells that was explored and rediscovered on countless occasions. Even symptoms could ultimately be reduced to intra-corporal phenomena: as Bourne remarked, symptoms were 'messages from a diseased area to the (patient's) brain' (Bourne 1931: 18). The problem for the investigating doctor was the line of communication after the signal left the patient's brain. Certainly, identification of the symptoms of disease involved consideration of what the patient said but if sounds were heard they represented at best the stumbling words of the disease. The interrogation was therefore concerned with:

> the characteristics and 'life history' of the symptom. ... To get a clear picture of the symptom so that it stands out as if it had a personality is the ideal to be sought for.
>
> (Stern 1933: 4)

To provoke the lesion to speech, through the patient, was not an easy task. Indeed, Keith, in his *Clinical Case-Taking* of 1918, while offering a basic schema for case-taking, pointed out that it was exceedingly difficult to reduce the skill of 'interrogatory method' to print (Keith 1915). Perhaps the commonest advice was that the patient should be 'allowed as far as possible to tell his story in his own words' when it came to identifying the main complaint (Hutchison and Rainy 1935: 2). The patient's own words were required in that they might express in purest form the communication of the pathological lesion itself. Such was the importance of these words that it was recommended that they should be written down verbatim (Gibson and Collier 1927).

Of course, it would be easy to project more recent sensibilities onto this interaction, perhaps to see the doctor as ignoring the 'whole person' of the patient, but that is to read the past through later eyes. As the texts on clinical method make evident, the doctor could only construe the patient as an anatomical body, as the physical density that surrounded the elusive lesion. True, this nineteenth century body had a separate identity – it was an individual apart from nature – but its individuality was circumscribed by the very techniques that gave it existence. Sanitary science only mapped a physical corporal space; and clinical medicine could only explore, analyse, and dissect that same volume, laying out to the view of the trained clinical eye, the detailed anatomy of its interior. Ironically, identity was more a characteristic of the pathological lesion than of the patient in that diseases had types that could be inferred through the signs (and symptoms) they engendered. It was therefore the voice of the disease rather than the voice of the patient that excited the interest of medicine in the early decades of the twentieth century.

Then, patient subjectivity began to emerge from below this threshold of description. Training the body to move and assembling a surrounding social space of interaction, meant that it was beginning to be possible to trace the outline of a sentient being. Just as movement had to be inserted into the inert density of the corporal body, so the ability to speak had to be interpolated into the awakening patient.

One of the first indications of an emerging new identity was the realization that the vicarious voice of the patient that spoke on behalf of the disease could be distorted. This communication problem marked the appearance of an idiosyncratic interlocutor in the traditional dialogue between disease and doctor as the latter began to recognize patients' different abilities to speak on behalf of the lesion. As Bourne noted in 1931: 'The human being is a recording instrument of uncertain and variable power' (Bourne 1931: 43). Thus, as a preliminary, the doctor had to assess the competence of the patient to verbalize the symptoms emanating from the pathology:

> As the patient describes his complaint, his mentality will become clearer, whether he is intelligent or dull, accurate or given to exaggeration, if his memory is good, or if there is evidence of mental aberration.
>
> (Gibson and Collier 1927: 7)

It was suggested that the interrogator should be alert to the possibilities of bias and formulate specific questions for each patient's intelligence (Noble Chamberlain 1938). Some texts introduced qualifications for the usual instruction to write down the patient's exact words: 'this does not mean that the observer is to set down words or phrases which are meaningless or equivocal' (Horder and Gow 1928: 1). Indeed an additional question was justified if the patient tended to stray into irrelevant matters. Some common words such as the patient's ready-made diagnosis were never to be taken at face value and, indeed, the interrogation could be postponed 'if the history is completely disjointed or the patient be all set for a three hour monologue' (Simpson 1937: 3).

Leading questions also distorted the patient's response and therefore posed similar problems. Hutchison and Rainy only allowed leading questions for stupid patients, to trap malingerers and to elicit subjective symptoms ('the morbid sensations experienced by a patient as the result of the disease of some organ or system') (Hutchison and Rainy 1935: 3). In other circumstances, leading questions were dangerous because of the 'suggestibility of many patients' (Simpson 1937: 3). If they were used, it was advised that a record should be kept and the patient's reply appropriately qualified.

Increasingly, the doctor had to look behind the words the patient spoke to differentiate those that should be ascribed to the lesion and those that emanated from this proto-subjective space. To be sure, the patient was still perceived as speaking on behalf of the pathology but the direct link between the disease and medical perception was splintered by the presence of this new shadowy intermediary. In effect, a new impediment had appeared in the space between doctor and lesion that disturbed the flow of communication; a small seed of subjective identity had appeared as the patient – naive, confused, suggestible – intruded into the great dialogue between medicine and pathology.

An inchoate patient

In the 12th edition of his *Physical diagnosis*, published in 1938, Cabot made two additions: he introduced a new first chapter on history-taking ('a subject too often omitted from books on diagnosis' (p. vii)) where before there had been none, and for the first time his chapter on the examination of the diseases of the nervous system included a two page discussion of the neuroses and the psychoses. Similarly, the eighth edition of Elmer and Rose's *Physical diagnosis* (revised by Walker) of 1940 added a chapter on history taking and a few pages on conducting a psychiatric examination despite their total lack in the 1938 edition (Elmer and Rose 1940). In the new schema, the old 'personal history', which had been more concerned with the patient's physical environment and habits, was replaced by an occupational history, a social history, which enquired after such personal experiences as worries, adjustments and disappointments, and a marital history ('domestic relationship, whether happy or unhappy, compatible or incompatible and the reasons for unpleasant relations, if they exist' (p. 24)).

The introduction of a 'mental' history into the consultation echoed concurrent changes in the field of psychological medicine. In the nineteenth century, when rationality had seemed all-important, psychiatry was concerned with those patients, few in number, who were insane. During the twentieth century, the central problem of mental functioning gradually became the more generalized one of 'coping' as medicine uncovered the wide prevalence of the neuroses – particularly anxiety and depression. By the 1930s, many doctors were well aware of the ubiquity of the neuroses and the need for a general mental hygiene. Consequently, patient anxieties and personalities together with notions of psychosomatic unity became important features of much advice on clinical practice and manuals on clinical methods began to include separate sections on conducting a psychiatric examination. In 1938, for example, Noble Chamberlain included a section on 'the diagnosis of the neuroses' in his chapter on the examination of the nervous system – though it was somewhat rudimentary (Noble Chamberlain 1938). He outlined the symptoms of the neuroses (hypochondria, neurasthenia, anxiety neuroses, compulsion neuroses and hysteria) but devoted most attention to their related physical signs: for example, he noted that in hysterics the ear lobes were ill-formed and were fused to the skin near the mastoid process instead of hanging freely.

Despite his new focus on mental health, Noble Chamberlain did not significantly alter his history-taking plan until the sixth edition of his book in 1957 when 'The home life' was introduced ('Is the patient happy and contented or are there sources of friction or worry?' (p. 6)). Even so, the 1938 edition did suggest a new goal for the history. In contrast to an earlier perspective that can be summarized in Horder and Gow's 1928 claim that they had 'in the main ... followed the well-proven principle of endeavouring to determine first the site of a lesion and then its probable nature' (Horder and Gow 1928: 2), Noble Chamberlain suggested that at the end of the history the physician should have:

> a mental picture not only of the patient's presenting symptoms, but of the manner in which these developed and of the background of personal and family life upon which they have been

> grafted. Too often we are rightly accused of studying the disease rather than the patient.
>
> (Noble Chamberlain 1938: 6)

This view of the patient signalled a significant change in the inferential logic of clinical practice. The relationship between sign and symptom that had dominated clinical thinking for a century began to be redrawn: the symptom was no longer always the subsidiary indicator of disease. For example, having observed that 'history-taking receives scant attention in other detailed books on physical examination' (p. vii), Bourne offered a discussion of the relative importance of history and physical examination and concluded that their significance in diagnosis or prognosis varied greatly with different diseases (Bourne 1931). Noble Chamberlain's 1938 text extended the argument with the observation that 'structural changes may exist without functional derangement and vice versa' (p. 2). In other words, the lesion might be unmarked by the sign so that the patient's words were not merely preliminaries but the primary access route to the medical problem.

A decade later this position was echoed by Cohen who, in the foreword for Seward's *Bedside Diagnosis* of 1949 poured derision on the student 'who diagnosed by structural resemblances' (Cohen 1949). Instead, he claimed that disease was 'a disturbance of function which may or may not be accomplished by structural changes' (Cohen 1949: v). Further, he contended that the traditional mind-body dichotomy was 'largely artificial' and that both psychosomatic disturbance and somatopsychic dysfunction were real phenomena.

Thus, by mid century it was possible for the symptom – the patient report – to achieve ascendancy over the sign. In the past, treatment success had been evaluated by the disappearance of signs; in the new medicine, the patient's attitudes were important. In 1947, for example, Dukes carried out the first survey of patients with permanent colostomy to see how they coped (Dukes 1948). His object was to examine patients' responses in the light of the operative technique as otherwise, he noted, there was no other means of establishing the best technique.

In the 7th edition of his teaching manual published in 1961, Noble Chamberlain added a new section and diagram in an attempt to show the complex relationship between signs and symptoms. He recognized for some diseases, in different stages of their development, that symptoms could be more important than signs.

Words and meanings

As well as re-assessing the relative importance of symptom and sign, the new cognitive map of medicine broadened the signifying focus of the symptom: while the sign remained firmly wedded to the lesion, the symptom detached itself and found a new target. Now the symptom could act as indicator both of the disease and also of facets of patient identity. When Hutchison replaced the patient's personal history with a social history in the 12th edition of his *Clinical Methods* of 1949 he believed this should include:

> the patient's mental attitudes to his life and work. ... One should endeavour to visualise the life of one's patient, sharing his emotions and viewing step by step his daily habits. ... Sometimes one should inquire into a patient's business affairs, his ambitions, anxieties, quarrels ... his domestic relationships, his psychological make-up, his interests, his hobbies, his hopes, his fears ...
>
> (Hutchison and Hunter 1949: 4)

An important component of disease still existed within the human body and this, as of old, demanded interrogation through the patient. There was now a second strand to medical perception, however, that identified a part of illness as existing in the shifting social spaces between bodies, and clinical method required techniques to map and monitor this space. The patient was beginning to have a voice – and an existence – independent of the pathological lesion; patients' words, as they described their symptoms, were no longer a vicarious gaze to the silent pathology within the body but the precise technique by which the new space of disease could be established; illness was being transformed from what was visible to what was heard.

The new meaningfulness of the patient's words was not, in this sense, a discovery or the product of some humanist enlightenment. It was a technique demanded by medicine to illuminate the dark spaces of the mind and social relationships. Whereas the pathological lesion could be visualized if it was given a neutral field, the illnesses of social spaces required the incitement of patient's subjectivity. At first, the patient was a fragile flower that had to be gently cultivated: as Hutchison noted in his 1949 edition 'one may defeat one's own ends by wounding the sentiments or conscience of the patient long before the physical examination starts' (Hutchison and Hunter 1949: 2). Later these new-born characteristics of personhood were to move to the centre of the medical project.

Perhaps one of the boldest attempts to provide a new meaning for the patient's words was that of Balint in the mid-1950s (Balint 1956). He argued that the traditional search for a localized pathological lesion was only a part – and often only a small part – of clinical practice. The role of the doctor, he suggested, was to organize unorganized illness: the doctor had to reorganize the patient's problems, symptoms and worries to make sense of them. This might require symptoms being linked with signs that together pointed to the pathological lesion in the classical triangulation method but it also required the placing of patients' words within a field of experience made up of feelings, symptoms and social context. Within this dense web of interconnections, the lesion was reduced to a single nodal point, no more important than other junctures in the network of relationships. Symptoms such as abdominal pain that had implied an abdominal lesion of some description could now equally well be linked with the patient's recent biographical events.

The reconstruction of patients' words from being a measure of medical effectiveness, to becoming the location of a major health problem in its own right, particularly through the notions of 'coping' and 'adjustment', began to take effect from the late 1960s, though its beginnings can be identified in the efflorescence of psychological medicine in the immediate post-war years. As the neuroses became common diagnoses, the doctor had even more reason to interrogate the mental space of the patient:

> Something more is required to establish this diagnosis (of the neuroses). It is necessary to assess the patient's personality, a task which comes more easily with age.
>
> (Noble Chamberlain 1952: 340)

Reconstruction of the meaning of symptoms had various implications for the patient's voice – and subsequent identity. First, it established a series of different needs that required expression. Thus, for example, Freidson, in 1961, in his *Patients' Views of Medical Practice* contended that 'performance of staff could not be understood very clearly without reference to the expectations of patients' (Freidson 1961: 12)) and in her study of patients and their general practitioner published in 1967 Cartwright could think of the effectiveness of general practice in terms of meeting patients' clinical, social and emotional needs (Cartwright 1967). Second, the patient's view was not only a part of the diagnostic process but also part of the therapeutic. At the very least, patient talk helped the process of organizing problems while, as a form of psychotherapy, it acted as a more formal treatment regime.

In the inter-war years the patient's view on anything, including the possible diagnosis, not directly related to the lesion was excluded; but in the second half of the century, it had changed. 'Patience is necessary', suggested Noble Chamberlain in 1967, 'when the patient tries to make his own diagnosis. This may be irritating but not unreasonable as it stems from a natural desire to find a cause for the illness which perhaps can be avoided in future' (Noble Chamberlain and Ogilvie 1967: 15). A decade later more than tolerance was required as it was possible, suggested Kleinman and his colleagues, that the patient's words – that were so coherent as to form explanatory models – could be valuable diagnostic and therapeutic tools in their own right (Kleinman *et al.* 1978). A flurry of work in the early 1980s on patient's 'lay theories' and of their importance for a penetrating medical perception, was further evidence of the elevation of the patient's own words from an irrelevance to a theory (Tuckett *et al.* 1985). The patient's words had been elicited in the 'interrogation', in the post-war years a less threatening term, 'history-taking', became more common while in later texts the even more secular 'medical interview' was employed. 'Clinicians are likely to consider the term "medical interview" as synonymous with which is called history-taking' wrote Enelow and Swisher in 1972; 'The medical interview is much broader than that' (Enelow and Swisher 1972: 3–4).

Medicine engaged with a new problem: the patient's words themselves. The patient's words were therefore more robust and the dangers of leading questions were minimized; by and large they should not be used but 'the student may observe an experienced clinician will sometimes disregard this rule' (Noble Chamberlain 1967: 16). Whereas before, the patient's words that did not signify the lesion were dismissed as irrelevant or suspected of representing malingering, the new advice was that it was 'important to realize that apparent evasiveness on the part of the patient is almost never deliberate' (Bomford *et al.* 1975: 3). The doctor's first task was 'to listen and to observe, not only to obtain information about the current problem but also to understand the patient as a person' (McLeod 1973: 1).

The space of the patient's identity revealed by medicine was not constituted simply by the words themselves as words were merely signifiers. The psycho-social world of the patient lay behind the words and further refinement of technique was necessary to make it accessible. At first, in the 1960s and 1970s, it was through an emphasis on non-verbal communication: 'the eyes sometimes convey more information than words; the clenched fist may demonstrate latent tension, and touch may be equally important' (Bomford *et al.* 1975: 3). By the late 1970s, the gaze beyond that which was spoken began to focus with more intensity on the subjectivity behind the words. And as this extended gaze grew in pervasiveness so subjectivity was consolidated. No patient could escape the enlightened techniques that demanded confession of emotions, ideas and experiences that, in their turn, began to constitute the main characteristics of an emerging identity.

The space of subjectivity

The 10th edition of *Clinical Methods*, published in 1935, had provided details of how to go about 'case-taking' (Hutchison and Rainy 1935). In the16th edition, published under new editors in 1975, 'case-taking' was not mentioned except in the chapter title and in its place the student was advised on how to 'take a history' (Bomford *et al.* 1975). At the same time it was never suggested that history-taking was other than a constant feature of clinical practice. 'History-taking is still an art', it was explained and it was '*a special form of the art of communication*. It is necessarily a two-way business' (p. 2; emphasis in original). Indeed, it was even suggested that history-taking might be improved by 'active participation in sensitivity groups as pioneered by (the GP-psychoanalyst) Balint' (Bomford *et al.* 1975: 2). In 1975, as in 1935, the patient was invited to speak while the doctor listened but it is clear that between these two dates the form of the invitation had changed fundamentally.

The interrogatory probe of clinical practice had entered the space of the patient's anatomical body in 1935 to read the nature of the lesion but by 1975, the measuring process of medicine had extended to embrace the subjective world of the patient. The unanswerable question of what the probe of 1975 would have found had it been used in 1935 might seem an intriguing one but it belongs to the politics of retrospective reconstruction that seeks to establish the continuity between past and present and to conceal the history of identity's emergence. To be sure, the interrogatory probe changed its form during the twentieth century but the probe was more than simply a tool of measurement: it was one of creativity. The clinical consultation with the patient was the productive machine of medical perception and practice. It was not simply the form of the incitement to speech but the very structure of perception that changed; equally, it was not what the patient said but what the doctor heard that established the reality (and accuracy) of the patient's world and identity.

The interface between doctor and patient was no longer the gap between an observing gaze and the anatomical space of the patient. The new medical perception interrogated the patient as a social space, as a body inhabiting a specific social milieu, but more important, as a space of subjectivity. The patient expressed their idiosyncratic self, their feelings, and their own experiences of the world. When Fitzpatrick and his colleagues published *The experience of illness* in 1984 (followed by a series of the same title), the patient was rendered as more than a corporal space, the patient had become an experiencing self (Fitzpatrick *et al.* 1984). Earlier Eisenberg had argued that illness and disease should be separated: disease could retain its nineteenth century links to the intra-corporal lesion but it should be separated from illness that reflected on the experience of the subjective patient (Eisenberg 1977).

During the twentieth century, the transformation of Man was profound. An anatomical identity had been extended through the assignment of movement and the constitution of a social space that opened up the strict limits of a corporal volume. That space of identity had now been incited to speak: at first it was a vicarious speech, the words of the intra-corporal lesion, then, gradually, it was a speech that revealed an inner world of experience located in the shifting corporal and social spaces of identity. The doctor's opening question 'What is your complaint?' was replaced by 'Now please tell me your trouble' (McLeod 1973). Illness, which had been bound to the intra-corporal disease lesion for over a century, broke free and made its new alliance with the idiosyncratic meanings of the patient's biographical integrity. And with this conceptual and practical gesture subjectivity was added to the expanding repertoire of identity. Those first stammered vicarious words had given way to a new fluency as Man had learned how to speak for himself and about himself.

8
Instilling Agency

Interrogation of the space of patient-hood gradually revealed a personal world of patients' thoughts, feelings and experiences – a psychological space existing somewhere between the corporal and social dimensions of identity, that could be located in either. A mind was ticking behind the exterior of Man, a mind that reacted to and reported on internal and external events; but could this psychological centre also be used to guide actions rather than simply record and report?

The potential of control

Sanitary science, like quarantine before it, had operated through a coordinated administrative structure to declare rules of hygiene and to maintain these within a legislative framework. However, there were limits to the effectiveness of this centrally controlled regime: laws could not easily be established to govern such personal activities as bathing, cleanliness or bowel movements. Certainly, where bodies came within the direct purview of the sanitary authorities – in the school, in the army, in the hospital, and in the prison – more rigorous schemes of personal hygiene could apply, but for the vast majority of the population the application of sanitary rules was less than comprehensive. How could hygiene be diffused more effectively throughout the mass of Man? The solution that emerged was a strategy that decentralized the management of hygiene enabling it to percolate through the population so that every person could begin to take some responsibility for monitoring their own bodies and boundaries.

The task of decentralization began in the early years of the twentieth century as experiments with exercise explored the extent to which control of the physical body could be transferred to the body itself. The shift from drill to physical exercise was only possible because some sort of decision capacity had been realized as lying behind movement. Whereas drill required an external authority and a posture that waited for the next command, physical exercise – particularly in its less formalized inter-war games and athletics – signalled the emergence of a centre of governance, of direction, of control, within that formerly inert corporal space of the nineteenth century. Exercise was therefore a body technique that not only replaced an idea of natural motion with purposive and directive movement but also served to give a sense of command to the body. It was surely no accident that physical exercise was held to build character and allow an outlet for the emotions.

The growing view that the body could manage its own movements and exercises was matched by the parallel emergence of psychological constructs that gave form to these basic impulses. Those nineteenth century 'natural' aspects of a proto-mental identity such as instincts, constitutions and habits could be translated into malleable characteristics of the individual mind: instincts could be moulded into attitudes and character into personality. The concept of behaviour, which ironically might refer more commonly to the style of motion of an inanimate object such as a ship, was applied to capture the contingencies of action, the purposeful movement of a self-conscious body. In effect, the management of body movement metamorphosed into the management of behaviour; behaviour was the label, the interpretation used to encapsulate a certain series of movements. Behaviour, however, even more than movement, was closely linked to mental functioning, at least in the new psychology that gave it life. The metamorphosis was most confidently expressed in the Freudian view that these 'natural' mental characteristics of Man were properties of the new-born infant rather than of some nineteenth century forebear: phylogeny was neatly collapsed into a myth of ontogeny.

The growth of inter-personal hygiene further transformed the capacity of the patient for action. Instead of relying on the vigilance of the sanitary authorities, the new public health began the process of recruiting patients themselves to the surveillance of body boundaries and anatomo-social spaces. Patients could be enlisted to practice their own hygienic regime; patients could become agents of medicine, their own self-practitioners. Thus, two iterative facets of the construction of behaviour were forged early in the twentieth century. On the one hand, patients as persons became more and more the objects of medical attention, particularly in their own actions and relationships. On the other hand, patients were also the subjects of medicine in the sense that they were recruited to monitor their own bodies. The active body of the patient required study, guidance and control if illness was to be avoided and health achieved. The malleable subjective mind of the patient in its turn demanded education and training if it was successfully to monitor and bring under some control its otherwise capricious body. The great dream of medicine began its strategic shift from the healthy body to the healthy behaviour.

Reactive behaviours

Despite the development of a framework in the first few decades of the twentieth century for handing responsibility for behaviour to individual patients, the patient as independent actor at first had only embryonic form. Indeed, the patient did not act so much as react. In one of the earliest health-related proto-behaviours to come to medical attention, defaulting from treatment (a particular problem in the medical management of venereal disease), autonomy was expressed by going against medical advice rather than by any pro-active behaviour. For example, in 1932, a paper in the *British Journal of Venereal Diseases*, reported on a survey of a group of defaulters from VD treatment (Frazer 1932). It concluded that it was important to 'impress on all patients at intervals the necessity of treatment' and that the patient should 'repeat all directions so no misunderstanding can occur' (Frazer 1932: 58).

Perception of the patient as a defaulter, or potential defaulter, represented a constant theme in mid-century discussion of the doctor-patient relationship. Yet within this overall concern with the problem of default there were gradual shifts as understanding of patients' movement/behaviour was redefined. One early development was an elaboration of the person behind defaulting, in particular a focus away from the problem of the generic defaulter towards the defaulter with individual characteristics. For example, in a series on default in the *British Journal of Venereal Diseases*, the defaulter was given a social status in papers on 'The Defaulting Seaman' (Hanschell 1935), 'The Defaulting Prostitute' (Nichol 1935a), 'The Defaulting Travelling Man' (1935b) and 'The Defaulting Child' (Nabarro 1935).

Continuing the attempt to locate defaulting 'behaviour' more fully within an individual context, MacFarlane and Johns reported in 1947 'a medico-social analysis' of 381 women patients with venereal disease (MacFarlane and Johns 1947). Then in 1948, in the first paper of its kind in the *British Journal of Venereal Diseases*, Wittkower discussed the 'psychological aspects of venereal disease' (Wittkower 1948). Following similar American studies, he had been asked by the War Office to investigate the personality, sex behaviour and 'driving forces' in a group of patients with venereal disease. He concluded by emphasizing the importance of psycho-social considerations in understanding the problem. Some two years later in a review of 'some individual and social factors in venereal disease' Sutherland declared that the social medicine approach must be applied to the study of venereal disease: 'the whole patient must be studied and treated as well as his infected tissues ... every patient is anxious and disturbed' (Sutherland 1950: 1).

In the post-war years, the problem of patients defaulting from treatment was reformulated using a more complex and subtle language of 'compliance' suggesting that obedience to medical instructions could not be assumed so starkly. The patient was thereby transformed from someone who simply failed to obey to someone who chose whether to follow advice. This new volitional component to patients implied that they could not be treated like passive bodies as in the past; the patient would require negotiation and persuasion if health-promoting ways were to be engendered. Thus, a new dimension was added to the doctor-patient encounter: good communication skills were becoming an important part of eliciting the patient's history but now they reached forward to influence the patient beyond the consultation.

It was assumed that patients' dissatisfaction and non-compliance were the product of poor communication (Ley 1976); equally problems of default and compliance became points on which to articulate the new concerns with 'effective' communication. At first, the task of communicating with the patient was framed in terms of giving reassurance. Cole felt that 'many people today carry unnecessary burdens of anxiety about their health which limit their happiness and activity' simply because they had not been given adequate reassurance about the doctor's prognosis (Cole 1946: 1). Armstrong too, in pointing our the neglect of the management and handling of patients in the medical curriculum over the previous fifty years, argued for the importance of allaying patients' anxieties through reassurance (Armstrong 1946). But reassurance was at best only an indirect way of influencing patients' behaviour, perhaps removing their inclination towards a 'sombre prognosis' with consequent undue apprehension, anxiety and fear (Parkinson 1951).

By 1960, it was becoming clear that communication with patients was something much more complex than the traditional concept of history-taking (Meares 1960). Verbal, extra-verbal and non-verbal channels of communication between doctor and patient were identified and explored. In 1963 a Sub-Committee of the Central Health Services Council of the British National Health Service produced a pamphlet entitled *Communication between Doctors, Nurses and Patients: an Aspect of Human Relations in the Hospital Service*. The Committee had been appointed in 1961 to improve information flow *to* the patient, though it concluded that little was known 'objectively' about patients and their reactions to treatment.

In his Rock Carling Lecture, 'Communication in Medicine', Fletcher took compliance as his central theme (Fletcher 1973). To be sure, that was a brief discussion on 'acquiring information' but the bulk of the review was on 'giving information'. 'How can we improve our communication with patients?' Fletcher asked. 'The first thing is to recognize the problem. Few doctors realize how little what they tell their patients may be understood or remembered' (Fletcher 1973: 17). In effect, behaviour was shifting from the innate waywardness of the patient, expressed perhaps in the form of their personality, to the contingency of the exchange between doctor and patient in the medical consultation.

Experiments in intervention

Since the nineteenth century, medical intervention had been targeted at the pathological lesion inside the body, but in the middle decades of the twentieth century a new locus for intervention opened up in the form of patients' movement/behaviour. Where better to target such aspects of patient identity but in those archetypal diseases of inter-personal hygiene that existed in corporal and social space, such as tuberculosis, venereal disease, problems of childhood and the neuroses? Later, the space of health behaviour would open up further so that after World War II a wider range of the population's movements could come under the eye of a medicine, but these later radiations out into the community were prefigured by two important inter-war experiments in Britain and the United States that demonstrated the practicality of monitoring and intervening in behaviour across a whole population.

The British innovation was the Pioneer Health Centre at Peckham in south London (Pearse and Crocker 1943). The Centre offered ambulatory health care to local families that chose to register – but the care placed special emphasis on continuous observation. From the design of its buildings that permitted clear lines of sight to its social club that facilitated silent observation of patients' spontaneous activity, every development within the Peckham Centre was a conscious attempt to make visible the web of human behaviour and interaction. Perhaps the Peckham 'key' summarizes the dream of this new surveillance apparatus. The key and its accompanying locks were designed (though never fully installed) to give access to the building and its facilities for each individual of every enrolled family. Besides giving freedom of entry, however, the key enabled a precise record of all movement within the building. 'Suppose the scientist should wish to know what individuals are using the swimming bath or consuming milk, the records made by the use of the key give him this information' (Pearse and Crocker 1943: 76–7).

The Peckham Experiment showed that behaviour could be monitored and its patterns revealed. From observation of patients' activities to the 'family consultation' at which health behaviours were discussed the Centre established a total overview of people's lives; but the crucial question of how to link the data produced by observing behaviour in the Centre with interventions to 'correct' such behaviour was never fully resolved. Whether or not patients were consuming milk or using the swimming bath did not carry direct implications for action: Should they consume less milk or more?

Should they use more sports facilities or less? And how was such behaviour change to be directed? The success of the Peckham Experiment was that it showed how the 'technical' problems of monitoring behaviour could be overcome. The next task was to see how the promotion of health-related behaviours could be integrated within a system of observation.

In 1923, the city of Fargo in North Dakota and the Commonwealth Fund embarked on a novel collaborative venture. The nominal objective of the project was the incorporation of child health services into the permanent programme of the health department and public school system and an essential component of this plan was the introduction of health education in Fargo's schools, supervised by Maud Brown. Brown's campaign was, she wrote, 'an attempt to secure the instant adoption by every child of a completely adequate program of health behaviour' (Brown 1929: 2).

Prior to 1923 the state had required that elements of personal hygiene be taught in Fargo's schools 'but there was no other deliberately planned link between the study of physical well-being and the realization of physical well-being' (Brown 1929: 7). The Commonwealth Fund project was a two-pronged strategy. While the classroom was the focus for a systematic campaign of health behaviour, a periodic medical and dental examination both justified and monitored the educational intervention. In effect, 'health teaching, health supervision and their effective coordination' were linked together; in Fargo, 'health teaching departed from the hygiene textbook, and after a vitalizing change, found its way back to the textbook' (Brown 1929: 7).

A preliminary look around the classroom 'disclosed a general pallor and listlessness' that led Brown to 'focus the attack on iron in the diet, fresh air, and sunshine' (p. 3). This, however, was more than therapy in an educational setting. From its insistence on four hours of physical exercises a day – two of them outdoors – to its concern with the mental maturation of the child, Fargo represented the realization of a new public health dream of surveillance in which everyone was brought into the corrective field of vision of the benevolent eye of medicine through the medicalization of everyday life.

Health behaviours

By mid-century, the importance of recruiting patients to their own care was fully recognized as health education became firmly established as an essential part of the public health programme. 'Health authorities', wrote Derryberry in 1945, 'are becoming increasingly aware that many diseases are uncontrollable without the active participation of the people themselves' (Derryberry 1945: 1401). Whereas a few decades earlier the community was only involved in public health in so far as their electoral or political support was required, now people had to be involved from the very beginning of any programme because they themselves were the agents of health practice (Derryberry 1949).

A major opportunity for the articulation of ideas on patient involvement in care emerged in April 1955 in the US when the Salk vaccine was declared safe and an effective protection against poliomyelitis. A national vaccination programme commenced at once and became the new focus in the public health literature on the problems of participation: here for once was a programme with undeniable benefits yet which the public failed to support wholeheartedly.

What were the public's attitudes towards vaccination? What factors precluded their involvement? Indeed, the questions began even before the vaccine was released. As Clausen and his colleagues noted in their report on parental reactions to the vaccine in the trials, they 'may be useful in indicating in somewhat greater detail than is usually available the factors influencing participation in such a program and the attitude and characteristics of participants and non-participants'(Clausen *et al.* 1954: 1526). It was not so much that the study of participation would clarify the problem of polio vaccination, but that the vaccination programme could be used to illuminate the whole issue of participation. 'Using the technics (sic) of social science research, it would seem fruitful to investigate further the methods of reaching and influencing those segments of the population which tend to be non-participants in such public health programs'(Clausen *et al.* 1954: 1536).

In similar fashion, response to community X-ray programmes for tuberculosis and to the new multi-phasic screening programmes focused attention on public participation (Hochbaum 1956). Yet how was participation and non-participation to be explained? The widely held view that as 100 per cent of Americans were of old stock, there were no cultural factors involved was wrong: 'obviously, culture is here as well as in Bali, if only we can see it around us' (Hochbaum 1956: 379). Thus, 'a person's motivations are compounded out of his previous personal experience and the cultural background from which he derives and in which he moves' (Wishik 1958: 139). Such attempts to explain behaviour in terms of culture and previous experience were formalized in the late 1950s when public health discovered beliefs as the mainspring of human action. The theories of Rosenstock and his colleagues of 1959, explaining why people failed to seek polio vaccination (Rosenstock *et al.* 1959), had become, by 1965, the basis for a more generalized model of health beliefs and focus on patient activity (Rosenstock 1965).

Again, patient involvement in the contemporary public health agenda provided the test-bed for understanding, managing and, ultimately, changing patients' behaviours to make them more congruent with the great dream of health. After the initial experimentation in the 1950s, the programme of targeting behaviour became increasingly a core component of the medical agenda as a generalized strategy of health promotion diffused throughout clinical practice. Concerns with diet and exercise, with smoking and dependencies, with stress and with sex, with anxieties and worries, with behaviours and body shape, might initially have been raised by medicine, but they rapidly became the vehicles for encouraging people to control their own behaviours in the name of health. These were also strategies that could usefully be deployed outside the strict confines of a medical care system: they could become the topics for the self-conscious development of 'lifestyle', for street conversations, for radio phone-in programmes, for newspapers and magazines, for celebrities and for ordinary people. The ultimate triumph of the new medicine of health promotion would be its internalization by all the population. A sense of agency, the ability to act with purpose, could be crystallized in every citizen around the small seed of a healthy life.

Using health services

One area of health-related behaviour that received special attention during the second half of the twentieth century was the use of health services. Such behaviour had been seen as unproblematic: patients experienced the symptoms of illness and then – almost as if they had been programmed to do so – sought medical advice. For example, when Horder and Horder carried out a survey of their patients' illnesses they found it 'difficult to believe' that two thirds went unreported to the health care services (Horder and Horder 1953: 184). How could this phenomenon be explained when a symptom or illness was a signal to seek medical help? What was clear was that such 'internal' behaviours could not adequately explain health service utilization; using the health service, like any other behaviour, had to be observed, dissected, understood, and then appropriate interventions devised to 'manage' the problem. Just as the patient's voice broke with its limited role as rapporteur for the lesion so patients' responses to illness had to detach themselves from the body's inner pathology.

In 1954, Koos published a book entitled *The Health of Regionsville: What the People Thought and Did About It* that was hailed as 'the first systematic explanation of what people think and why they behave as they do in regards to health' (Koos 1954). Koos and his colleagues interviewed more than 500 families over a 4-year period in an American town, which they gave the name Regionsville. Respondents received some 16 interviews, each one with a different focus. For their views on illness, patients were provided with a checklist of 17 'readily recognizable symptoms' and asked which they thought should be brought to the attention of the doctor. Those who reported no disabling illness in the previous 12 months were asked whether they had experienced any of the symptoms from the list.

Analysis of 'what the people thought and did about it' marked the beginning of an increasing interest in patients' words by the human sciences during the ensuing decades which almost exactly paralleled the growth of the requirement for an 'extended history' in medicine, but it also reflected the growing interest in health-related behaviours. Identifying how patients would respond to the experience of a hypothetical symptom became a popular technique for social surveys during the 1960s and 1970s. For example, in 1960 Apple published a study entitled 'How laymen define illness' in which 60 respondents were given eight descriptions of people with health problems and asking: 'Are these people sick?', 'What might the illness be?' and 'What should be done about it?' By varying the degree of ambiguity and time of onset of the problem, Apple was able to show the significance of various aspects of an illness for promoting patient action (Apple 1960). Similarly in 1961 Baumann invited 201 patients (and 262 medical students, acting as controls) to answer the question 'What do you think most people mean when they say they are in very good physical condition?' The replies were subjected to 'content analysis' to establish three elements: general feeling of well-being, absence of general or specific symptoms of illness and what a fit person would be able to do (Baumann 1961).

In 1960 Mechanic and Volkart explored the propensity to visit the doctor by providing 614 college students with a checklist of symptoms and asking them whether they would take them to the university health service ('certainly, probably, not very likely or very likely') if they had them (Mechanic and Volkart 1960). From the varied responses, they derived the concept of 'illness behaviour' to explain why some patients did and other patients did not visit the doctor with a given symptom. The concept of illness behaviour was part of the post-war fascination with the weakening person-patient interface that emphasized again and again that the transition between health and illness, between person and patient was not predicated on an absolute and biological difference. Rather, it was underpinned by the notion of a person as their own health practitioner making their own judgements and decisions on the nature and boundaries of health and illness, and acting accordingly.

Perhaps the date for the beginning of this transformation of patienthood can be established from the sudden interest in the placebo effect from about 1948. Before the war, the placebo effect went unnoticed: how could it possibly be identified in patients who functioned as passive bodies? Recognition of the placebo effect signalled the moment when subjective mental processes began to seize control of the body. Patient thought and deed, both conscious and often unconscious, had an effect on health. Medicine invented the placebo effect to describe this new patient role; it also discovered the doctor-patient relationship as a new and tendentious topic that demanded debate and analysis. The relationship was between practitioner and body as of old, but also between doctor and patient in a social context (Parsons 1951), between physician and subjective mind (Balint 1956) and between medical practitioner and quasi-medical practitioner; the doctor consulted as much with his patient as an 'agent' as with his patient as a body; the patient was an 'expert' whose actions were informed by lay medical knowledge and theories (Tuckett *et al.* 1985).

In 1900, medicine was largely represented by the physician armed with tools for the exploration of individual physical bodies; the patient was coterminous with that body, essentially passive, involved only to the extent that it must respond to and report the outward manifestations of the organic disease. Health was the state that was gained or lost by the disappearance or re-appearance of the lesion. In the latter half of the twentieth century, however, the relationship between medicine and the patient became more complex. The patient became a 'consumer' of health care, an object of the medical enterprise, a psycho-social being whose every movement and gesture was monitored, evaluated and re-ordered by a medical machine based on a new ideology of health. In parallel, the patient became a 'producer' of health through health-protective behaviour (Harris and Guten 1979), a self-practitioner engaging in self-care. As both consumer and producer, the identity of the patient was transformed: from the passive object of the nineteenth century contained within strictly anatomical boundaries to a behaving person within biographical and social space.

In the early twentieth century, the encouragement of exercise and inter-personal hygiene provided two important facets of the new inter-corporal space that was opening up. In the second half of the century, these elements came together in an emphasis on the behaviours that might secure the individual's health. Then the moving participating individual proceeded from behaving to acting – better to capture the voluntaristic element in human activity; illness moved further from the body of the patient to social spaces, to problems of coping and adjustment; and medical strategies shifted from the pronouncements of experts towards 'informed choice', self-efficacy, non-directive counselling and non-judgmental health behaviour as the old referents disappeared. Thus, alongside the incitement of subjectivity there was the invocation of reflective thought: not only thought of the subject but also subjective thought about self. When George Bernard Shaw chose the classical story of Pygmalion as the basis for his celebrated stage play of 1913 the story of a sculpture that came to life provided an inexact metaphor for the struggle to educate Eliza Doolittle. Professor Higgins did not try to transform her body but her self-awareness and behaviour; that was the challenge for every-body in the twentieth century.

The emergence of reflexivity represented the culmination of a series of attempts to construct a notion of agency within the space of the nineteenth century corporal creation. Identity was changing and was changing itself. The Darwinian solution – ironically evolving with great rapidity to encompass the transforming figure of Man – was to anchor these changes in a eugenic utopia in which Man was empowered but with the goal of returning to a nineteenth century model of anatomical perfection and stability. And what a price for defective bodies that dream entailed. Yet, it was all too late; agency had already escaped its biological confines. The new space was not corporeal in nature, nor was it exactly inter-subjective. In the sense that it involved reflection about self it was intra-subjective; but this characteristic depends on the exact location of the subjective. Was identity coterminous with the body? Was identity crystallized in interpersonal spaces? Or did reflexivity unfold across yet another series of spatial axes that defined a new ontological realm in which the project of Man could be realized?

Concerns with health behaviours and their origins in the patient's mental processes was matched by contemporary anxieties about control over behaviour. These were expressed formally in notions such as 'locus of control' (Rotter 1966) particularly in relation to health but had their precursors in Cold War paranoia. Could US prisoners in Korea be brainwashed to change their behaviour? Could a 'red under the bed' of McCarthyite America really be a threat to the innocent citizen? And could aliens (a contemporary creation) pose a threat to the minds of ordinary citizens by taking over their bodies as portrayed in a whole genre of post-war science-fiction films? The images were frightening but the 'nightmare' did happen: minds were taken over; the autonomous mind of Man woke from its pre-programmed slumber and took control of its own behaviour. Poor Frankenstein's monster, he did not understand the world in which he was born, but instilled with the power of agency he could at last master his uncontrollability and become the self-aware, self-directing citizen of tomorrow.

9 Confessing Death

In the first half of the twentieth century mechanisms for promoting an anatomical identity were increasingly complemented by processes that monitored inter-personal space and incited subjectivity in the medical consultation – with the consequent appearance of a non-corporal psycho-social identity. Even so, the nineteenth century idea of pathological death that provided one of the central underpinnings for the biological identity of Man continued into the twentieth century as the independence of corporal space was reaffirmed at all stages of its existence. By mid-century therefore, there was an increasing tension between a death that provided the basis for an anatomical self and those 'humanizing' procedures that invoked a subjective identity and sense of personal agency. The tension was resolved by the construction of a new form of death and dying.

The secret and the lie

In the second half of the twentieth century, new ways of thinking about death began quite rapidly to usurp the old notion of pathological death that had provided a way of knowing about and analyzing death for the previous one hundred years. This was not a process of gradual discovery, of truth being revealed having been hidden for a century or more. This was the replacement of one system of understanding by another, the substitution of one regime of truth by another. The new regime was not in any fundamental way 'better', it simply addressed and answered a different set of questions that had emerged as the clinician in the consultation coaxed out the subjective world of the patient.

The collapse of one regime of truth and the emergence of the new one occurred over several years but the key moment of transition can be identified through changes in what could and could not be said. Just as pathological death had produced a new language for analyzing the end of life, so the new form of dying introduced new ways of speaking and hearing. In general, this transformation can be represented as a change in the dominant regime of truth, as what was considered as true and what was considered as a lie at any point in time.

The lie about death first made its appearance in the 1950s when a new debate arose on the question of whether or not to tell the patient of the imminence of death (Standard and Nathan 1955). For about one hundred years, the nature of death had been revealed in the post-mortem room: now, medicine began to debate whether or not patients should be told that they were dying. Should patients be told the truth? Or was the 'true' way of dealing with the problem to tell a lie? This new debate could not be resolved by empirical enquiry as it was not a question of what did happen but rather what should happen. Indeed, the very existence of the debate suggested a weakening in the established ordering of death, the beginning of the moment of transition, the moment when one system of knowing began to be destabilized and the next had yet to take shape. When it is debated whether truth is to tell or not to tell then clearly truth has no precision; and when it is unclear whether truth demanded certain words or others then the boundary that marked out the lie had not yet fully formed. There was even a point in this confusing transition when the liar was one who told the truth to the patient.

The redrawing of the line between the truth and the lie can be illustrated by the discovery of the secret. A secret represented the truth that could not be told. To keep death a secret was justifiable because patients inevitably feared death and relied on the hope that the secret gave them: 'optimism is the greatest analgesic. Hope is the most certain tranquilizer' (Ogilvie 1957: 591). Accordingly, it gave the patient no benefit to be told that they were dying (Asher 1955). The secret could not be spoken as that would destroy it, but neither was it a lie. Indeed the secret could, in mysterious ways, be passed between doctor and patient without speech: 'Only by an understanding look or a long squeeze of the hand is the secret that might be unendurable sometimes communicated' (Ogilvie 1957: 591).

Defence or justification for not telling the patient rested on whether or not patients had a wish or right to know. Thus between, say, an observation in 1957 that 'as a rule ... patients do not ask, or if they do they do not want the truth, but were only seeking to be reassured' (Ogilvie 1957: 591), one in 1959 that the patient might be 'told' but not if it 'might induce psychopathology ... or destroy hope' (Aronson 1959: 253), and one in 1963 that 'most patients wish to talk of their situation and they usually experience relief when given the opportunity of a frank discussion' (*Lancet* editorial 1963: 927) lies a great discontinuity as one understanding of dying replaced another.

As the problem of what to 'tell' changed in the late 1950s and early 1960s so the secret became 'the most dreadful question of all' (*British Medical Journal* editorial 1963). Then it was exposed as the lie. Death was seen as being surrounded by a 'conspiracy of silence' (Platt 1963). In contemporary analyses, the encounter between doctor and patient could take various forms, each dominated by secrets and silence: there was 'closed awareness' in which the medical attendants chose not to tell the secret, 'suspected awareness' in which the patient had a hint of the truth, and 'mutual pretence' in which both parties knew but both chose not to share their knowledge (Glaser and Strauss 1965).

The secret was a bond between physician and patient that prohibited its discussion. If the secret was tentatively to be explored it would have to be in marginal places where concealment held less sway. Thus, while it might be dangerous to speak to the dying it was possible to ask the mentally ill about their attitude to death (Bromberg and Schilder 1936; Caprio 1946; Feifel 1955) because their disturbed understanding prevented knowledge of the secret. Next in line to join the new discussion of death was the patient's family whose wishes had to be established (Alvarez 1952), and then the medical management of dying required the family's view of death to be further articulated, explored and discussed (Aitken-Swann 1959). Finally, as the secret was transformed into the lie, it became time to involve the patient. In 1959, Feifel added 85 'normals' to his sample of 85 mentally ill and 40 'older' people when he asked: 'What does death mean to you?' (Feifel 1959).

At first, the lie had been sustained by hope, and then by patients' fear, which in its turn was reinterpreted as distress produced by the same lie (Hinton 1963). From silence to speech, a new imperative emerged: 'The greatest need is for a listener who will try to understand and help to relieve the patient's sense of loneliness and deprivation'(*Lancet* editorial 1963: 927). Even silence – that had once existed as the antithesis of speech – was joined with its old opposite in a new more penetrating form of communication: 'those who have the strength and love to sit with a dying patient in the *silence that goes beyond words* will know that this moment is neither frightening nor painful' (emphasis in original) (Kubler-Ross 1969: 321).

This new-found analysis of the process of dying found its clearest expression in the notion of anticipatory grief. Anticipatory grief, which had been 'customarily applied primarily to prospective survivors, may be applied to the prospective deceased as well' (Aldrich 1963: 331). Since the beginning of recorded time relatives and friends had led the mourning procession; now in a great reversal the chief mourners became the dying themselves. Thereafter truth was embodied in the words of the dying patient as the secret was broken and the confession exacted.

It would be too simple to see the change in terms of a scenario that held that previously the doctor had withheld information ('the lie') and now told it ('the truth'). Indeed, it was reported as rare for the doctor to 'tell' the patient. Telling assumed the patient could hear and the doctor 'knew'; but it was less a question of what was said than how one went about speaking (Hinton 1963). With encouragement and interrogation, the patient came to realize and admit to their own deaths.

Unlike the dialogue of the early nineteenth century, the new discourse was not between the dying and their friends and relatives but between the dying and a doctor, nurse, almoner or chaplain (*Lancet* editorial 1963: 927). Moreover, while, as of old, the domestic setting may have been the most conducive to encourage the patient to speech, this could be reproduced in the informal yet medicalized space of the hospice (Davidson 1978). Moreover, the mechanism for transmission of the secret had no need to rely on lay relationships or, for that matter, on the understanding look or the squeeze of the hand, but on the increasingly deployed techniques of counselling through which the secret was shared.

Recognition of the 'stages of dying' (Kubler-Ross 1969: 321) ensured that whatever the mental state of the patient, whatever their points of resistance in terms of denial, withdrawal, anger, and depression, they would in time be guided to realize and share the secret of death. Denial moved from being an attribute of the secret to a 'manifestation of the therapeutic gyroscope of the psyche' (Schneidman 1976: 240). Patient denial demanded neither challenge nor complicity but investigation, an exploration of a diagnosis of not wanting to known (Wiesman 1972). The death-bed confession of the sinner and the guilty, for so long a ritual purification of the soul by the church or a melodramatic device in cheap literature, within a few years became the enlightened regime designed for a whole population as innocent and guilty alike had the truth wrested from them and their confession heard.

The general deployment of pathological death during the second half of the nineteenth century and first half of the twentieth, celebrated the integrity of the body and its separateness from natural forces. With the late twentieth century twist in the conceptual framework that informed death and in the many practices that surrounded it, the focus was not the autonomy of corporal space but of psychological, subjective space. And with the invitation to reflection about self-knowledge in the talk of the dying the final movement into reflexivity was achieved. The flat inert body that medicine had previously treated so passively had to be incited to speak. In many ways, it echoed the encouragement given to patients to speak during the consultation: just as the patient's subjective autonomy was being revealed by urging speech about an inner life, the same technique could be applied to death and dying by placing the patient in confessional mode. A pathological death had celebrated an anatomical body separated from nature; now the new confessional death of the years after around 1961 ensured that the patient broke out of the corporal chrysalis and shared with others (in the form of their clinical attendants) the inner-most secrets of their subjective selves.

The biology of death

With the advent of a confessional death, the older pathological death began to lose much of its meaning and, more importantly, its central role in determining the truth about death. Pathological death had held its grip from the middle of the nineteenth century by means of the everyday ritual of completion and processing of the death certificate. Certificates were filled in as an everyday routine; then they were forwarded to a central office where they were collated and analysed. Finally, the results were published as mortality statistics. Death was thereby translated from its role of defining a body to one of characterizing a population.

Mortality statistics acted as a point of articulation for various shifts in perception of the nature of bodies, health and populations. For example, mortality rates provided an index of the standard of healthy living of the population, or an assessment of the effectiveness of medical intervention and of health care provision. This is not to say the production of both the death certificate and mortality statistics, although routinized, was entirely without its problems. It was known, for instance, that many medical practitioners used outmoded or inappropriate terms to describe cause of death. In addition, the classification system itself that categorized the individual death certificate was subject to decennial international revision as it strove for greater accuracy. But perhaps the greatest difficulty lay in handling multiple causes of death. This latter problem had first emerged at the beginning of the twentieth century when the statistical notion of correlation began to replace causality.

A cause was localisable to a specific point in human tissue whereas a correlation was more a mathematical abstraction that made it possible 'to go on increasing the number of contributory causes' (Treloar 1956: 1377). Numerous revisions of the format of death certificates occurred during the early decades of the twentieth century, each one attempting to capture the distinctiveness of primary and secondary, immediate and distal, contributory and non-contributory causes of death. Which 'cause' actually brought about death? Which 'cause' was clinically significant? Various rules and routines were devised to enable classification coders to assign the more appropriate cause with often confusing results so that eventually it was left to the certifying doctor. In effect, while the ritual of death certification was refined during the nineteenth century so that it might more accurately express the cause of death, by the late inter-war years of the twentieth century the certainty that the certificate and mortality statistics spoke the truth about death began to look more uncertain. By the 1950s, it had become clearly:

> a false proposition that each death is a response to a single cause. ... Efforts in recent years to solve the problems of tabulation and interpretation of assignment of multiple causes represent an attack on a basic problem and clearly indicate that causation of death is not considered singular.
>
> (Treloar 1956: 1376)

The crisis of confidence in the traditional objectivity of the death certificate was also apparent in the debate concerning the accuracy of cause of death recorded on the certificate. There had been challenges to the accuracy of death certificates earlier in the twentieth century but these concerned primarily the over-recording or under-recording of certain diseases as doctor's diagnostic practices failed to keep pace with changes in the nomenclature and classification of disease. In the 1950s and 1960s, however, the overall accuracy of the many processes of death certification came increasingly into dispute.

For one hundred years, the autopsy had functioned as the temple of truth. There, laid out in the depths of the corpse, was the cause of death for all to see. The existence of the lesion that caused death had in most cases already been inferred by the attending clinician and the autopsy, when it was held, was merely confirmation of the diagnosis. Yet a variety of studies in the 1950s showed large discrepancies between the diagnoses of clinicians at the bedside of the dying and of the pathologist at the autopsy, sufficient at least 'to show the need for care in the interpretation of mortality data and the difficulties of arriving at an accurate diagnosis of cause of death' (Heasman 1962: 733).

From representing the solid anchorage of medical practice, the pathological lesion that 'caused' death became equivocal. Mortality statistics had been 'hard ... but several studies have shown how inaccurate the actual entries on death certificates often are' (*Lancet* editorial 1970: 1072). What had been an unquestioned assumption became contentious: 'Until recently the epidemiological literature has contained many analyses of recorded mortality data but little concerning their reliability' (Florey *et al.* 1969: 15). Even the solid referent of the autopsy report came into doubt. What value 'cardiac arrest' as a cause of death when it was 'how all of us will leave the world whatever the real cause of our death' (Medical Services Study Group 1978: 1065)? 'In what proportion of cases do we really know the cause of death? In a very large proportion of necropsies some anatomical lesion is found, but assessing its importance is more a matter of philosophy than fact' (Emery 1962: 740).

Consistency in certification between clinician and pathologist, traditionally the bench-mark of validity, was only to a minor degree a test of accuracy of diagnosis, reflecting more on consensus: 'It is biassed in favour of a persistent diagnosis' (Emery 1962: 740). Death records simply represented 'the relative ease with which we can secure unanimity of agreement in selecting one item out of a causal system as being the important one for the purpose in view' (Treloar 1956: 1376). In the past, the death certificate and the mortality statistic had represented immutable fact, the truth of death and disease; in the 1950s and 1960s, they began to express a different sort of meaning.

Death had been the concern of pathologists, but they began to lose their hold: 'We are mistaken to consider death as a purely biologic event' (Feifel 1959: 128). On the one hand, the pathologist's analysis of death was criticized for failing to represent accurately the biological reality of death, but in addition, it was also suggested that for some categories of death, particularly suicide, the death certificate should record non-biological data such as the intention of the decedent. This would require enquiries of the immediate family and friends about the patient's mental state prior to death, a sort of psychological autopsy. The dream of the new investigative framework involved a total autopsy that would include 'the services of the behavioural scientist, psychologist, psychiatrist, sociologist, social worker' (Schneidman 1976: 248).

With the new confessional death, the justification for euthanasia was inverted. For much of the twentieth century euthanasia had been a component of eugenics, a procedure advocated by others for those who were in some way sub-human. In 'voluntary' euthanasia, it became a component of a discourse that gave the patient the right to speech and a claim thereby to have their human-ness respected (Downing 1969).

In summary, in the immediate two post-war decades death certificates and mortality records moved from forming the hard bed-rock of medicine to combining subjective impression, arbitrary rule and professional consensus. Death certificates no longer spoke a biological truth about the body but became a ritual activity: they 'are not primarily intended for epidemiological research they are an important legal and social requirement' (Medical Services Study Group 1978: 1065). In principle, death certificates should have helped to define the problem of morbidity but 'certainly no statistician worthy of his salt is going to accept even the best mortality records as other than a grossly biased sample of morbidity conditions in a total community' (Treloar 1956: 1376). Contemporary interest in measuring morbidity more directly, together with increasing focus on those illnesses particularly chronic that had not produced a mortality statistic, provided further evidence for the shift in medical perception.

Analytic space

From the mid nineteenth century, the analytic space on which medicine focused was the biological realm of the human body. It was the body that had to be scrutinized for the secrets of death, the body that had to be sanitized and guarded to prevent death from contaminating the living; it was the body that had to be made a public object through the compilation of statistics on its birth, death and cause of departure. This biological space was the repository of truth; beyond there was nothing and therefore no need to investigate the communications that passed between the dying and their entourage.

About 1960 the diagram of knowledge was rearranged. The human body was no longer the dominant space of truth that required analysis and interrogation. The cause of death was fractured, each sign, each finding deep in the body that previously had spoken the truth about the cause of death, was now of ambiguous meaning. The death of the body could no longer be analysed as a singular event in which truth was captured; death was a temporal trajectory of dying (Glaser and Strauss 1968). Psychologically death was a process in which the dying negotiated stages and the old might suffer pre-death (Isaacs *et al.* 1971). Biologically, death was dispersed as the interrogation of the body led to somatic death, organ death, molecular death, etc.; indeed, there were as many deaths as tissues and cells. From the early 1960s when the flat electroencephalogram that was held to mark brain death came to dominate analysis of the moment of death (Hamlin 1964), the time of death became a matter for debate. For the international Declaration of Sydney of 1968, irreversibility of the processes leading to death was identified as the key event (*British Medical Journal* editorial 1968). Taken further, this meant that death now encompassed life: dying was a 'long-drawn-out process that begins when life itself begins and is not completed in any given organism until the last cell ceases to convert energy' (Morison 1971: 695). Dying and living had become coterminous; dying had become a sort of living (Kubler-Ross 1981; Saunders and Baines 1983). Biological death became more the province of the medical ethicist than of the clinician as a new philosophical analysis arose to usurp the domain of pathology.

It was not that the medical analysis that had radiated through the body became silent: it simply addressed another space. In part, this new space was one in which biomedical interest moved from the fixed structures of pathology to the constantly shifting processes of life. Yet this discourse, particularly in the hands of the medical ethicist, was further revised to focus not on the body but on the expert – the clinician, the pathologist, the psychologist – who spoke. The problem of inaccurate or unidentifiable death became a problem for those who created the inaccuracy or those who were unable to identify the moment of death. The secret of death and the truth of life no longer resided with such assuredness in the depths of the body; the court of judgement demanded less the body as evidence and more the person as witness. Thus, the new discipline of medical ethics took over the medical analysis of death and required the confession of the clinician while the new discourse on dying encouraged the dying subject to speak. For one hundred years, the body had spoken; now it was the reflexive subject, the new 'self'. It was therefore not surprising that during the twentieth century cremation replaced the elaborate sanitary rituals of interment for disposal of the dead.

As in the nineteenth century, reconstruction of death went hand in hand with a reconstruction of life. In the nineteenth century, a movement from natural to pathological death marked the separation of the body from nature and its constitution as an object in which could be instilled movement, behaviour and subjectivity. Late in the twentieth century, a revolution of equal importance replaced pathological death with confessional death. The shift was portentous in many different ways. The corporal body that had so dominated individuality for a hundred years was collapsing, the secrets of death were no longer to be discovered in the anatomical depths of the human body; instead the secrets of life were to be discovered in the reflective words of the dying. Just as the consultation had invited the patient to speak about who they really were, so the confessional of the dying produced a brief efflorescence of reflexive identity.

Historical reconstruction

Spread of the new confessional death was extensive, just as were the crises in the old pathological form. What remained was the debris of the transition as new replaced old. The residuum, in its turn, had to be placed within an explanatory framework that could account for the demise of the old regime. Like the body, the old truth had to be decently interred. The solution was to use the persuasive devices of liberation and enlightenment to explain the triumph of the new. Aries' history of death, published in 1981, provides an exemplar of the techniques that could be deployed (Aries 1981).

The problem was how to account for three regimes of truth surrounding death, roughly that preceding 1850, that between 1850 and 1950, and that after 1950. The solution was to romanticize the earlier period, deplore the intervening century, and acclaim the recent changes. Thus, Aries contended that between about 1850 and 1950 there occurred 'a complete reversal of customs' involving 'the beginning of the lie' in which the dying patient was no longer told their prognosis and 'death is driven into secrecy' (p. 562). It involved the medicalization of death such that the dying were removed to the sanitized space of the hospital to conceal the new indecency of death; and it involved the privatization of death through the rejection and elimination of ceremony, ritual and public mourning:

> Once, there were codes for all occasions, codes for revealing to others feelings that were generally unexpressed, codes for courting, for giving birth, for dying, for consoling the bereaved. These codes no longer exist. They disappeared in the late nineteenth and twentieth centuries.
>
> (Aries 1981: 579)

In this period, Aries suggested that he detected 'a return of the hideous images of the era of the macabre' when death 'turns the stomach like any nauseating spectacle' together with the impropriety of 'the biological acts of man, the secretions of the human body ... the smells of sweat, urine and gangrene' (Aries 1981: 579). It also marked 'the completion of the psychological mechanism that removed death from society, eliminated its character of public ceremony, and made it a private act' (Aries 1981: 575). In the domain of silence that Aries claimed marked the medicalization of death, the hospital became:

> the place of the solitary death ... the only place where death is sure of escaping a visibility or what remains of it that is hereafter regarded as unsuitable or morbid.
>
> (Aries 1981: 571)

The apparent regime of silence and denial that began in the mid nineteenth century reached its greatest hold at the end of the 1950s, but then, almost as suddenly as it started, it would appear to have weakened, especially in the Anglo-Saxon world.

> A complete reversal of attitudes has been taking place (promoted by) a group of psychologists, sociologists and psychiatrists who became aware of the pitiful situation of the dying and decided to defy the taboo.
>
> (Aries 1981: 589)

According to Aries, the new regime of truth-telling humanized death, releasing it from the repressive regime that had held it so tightly in its grip for the preceding one hundred years. It marked a return to those days early in the nineteenth century when, apparently, death was not surrounded by lies and the truth could be spoken freely. It represented the abolition of an enforced regime of silence that prevented the dying saying what they yearned to say.

Thus, by 1981, a bare two decades, after the revolution in the conceptualization and management of death, the circle was squared. Not only had one regime of truth replaced another but also the process had been 'explained' in terms of progress and enlightenment. Yet perhaps more importantly, this historical reconstruction held that Man always wanted, nay, needed to speak the truth about himself but was prevented from doing so between 1850 and 1950. The reassessment was therefore more than a 'progressive' gloss on events. The reconstruction affirmed the new identity that had emerged in the mid twentieth century and gave it universal status. Not only was Man a subjective and reflexive being from the moment of Aries text in 1981, but also, with a bold and retrospective sweep, always had been.

Aries' fable supported well the dominant Darwinian creation story that placed Man's origins in a long distant past. In Darwin's own account it had been anatomical identity that had been forged in that evolutionary past, but this explanatory framework could just as easily carry the burden of the new transformed and transforming subjectivity. The recent advent of subjectivity could be projected back in time to establish a subjective (and reflexive) self from its point of origin. Like a shadow play, the evolutionary past of Man could be reconstructed alongside the parallel construction of Man himself. Indeed, this project became the centre-piece of the New Darwinism as it sought to extend the Darwin creation story to encompass those psychological and social traits that had emerged so recently. Despite his emphasis on the psychological rather than the biological there is really very little between Aries' revisionist history of death of 1981 and Wilson's earlier polemics on socio-biology of 1975 or Dawkin's *The selfish gene* of 1976. Each analysis attempted to prolong the once revolutionary nineteenth century creation story by extending its remit to embrace changing identity. With each explanatory success the immanent constancy of Man since before history was affirmed and reaffirmed. Like the omnipresence of God's work that sustained the Biblical story of creation for more than a thousand years, the Darwinian story was revised and enhanced to accommodate its changing and dynamic object. And with its totalitarian hold the shifting spaces and moving geometry of a continent of identity was lost in a hall of mirrors.

10
Dimensionalizing Identity

The main classification principle in nineteenth century medicine had been a binary one. This duality had been expressed in the way that sanitary science separated corporal and non-corporal space and in the procedures of clinical medicine that identified worlds of physiology and pathology, the struggle between life and death, the difference between being healthy while life forces maintained their superiority and being ill as disease gained the upper hand. The realignment of the space of the body during the twentieth century, however, meant that it was not juxtaposition with nature that provided the basic comparator for identity but the relationship with other bodies. Inter-corporal comparison did not imply a binary classification but a multi-dimensional one in which one body was held in relation to the many characteristics of many other bodies. The development of this new space of identity is most clearly shown in the rise of the idea of the normal during the twentieth century.

The normal child

No doubt there were nineteenth century manifestations of the idea that a person – or more frequently, a population – hung precariously between health and illness (such as attempts to control the health of prostitutes near military establishments), but it was the child in the twentieth century who became the first target of the concerted attempt to conflate normality and abnormality and thereby create a scale of difference. Ironically, the body of the child had emerged late, being wrested from nature several decades after the original body of Man. In part, the problem had been the strength of the infant's affinity with nature: it required more than the policing of a boundary to pull its small body out of the grip of tenacious natural forces. And not only did it involve a lasting separation from nature, but also – as if the wrench was so great – the infant was propelled into a new social space of safety.

The space that surrounded the new-born infant was a multidimensional one. It was, of course, definitively 'non-nature', yet in addition, it was a space dissected by analyses of nutrition, domesticity, legitimacy and social class, insinuating that these same dimensions characterized the identity of the child. Even more important, it was also a space traversed by growth and development such that there was a constant threat that proper stages might not be negotiated. This meant that the development of the child had repercussions far beyond their individualities. The potential was enormous. In public elementary schools of England and Wales, there were five million children:

> a population which is congregated and readily available for training in citizenship or for any health or other social improvement schemes of the local educational authority; a population which opens the door to contact with parents and gives a rough guide to home conditions; a population which is under a measure of discipline; a population which is impressionable, receptive, plastic, and pliable; a population which under sympathetic and proper guidance should be in the process of developing habits which would conduce to the welfare of the individual and the community.
>
> (Lloyd 1936: 207)

The establishment and wide provision of antenatal care, baby clinics, milk depots, infant welfare clinics, day nurseries, health visiting and nursery schools enabled the early years of child development to be closely monitored (McCreary 1936). Ensuring that the developmental stages of childhood were successfully negotiated also produced a close alliance between education and health care with inspection clinics that screened all schoolchildren at varying times for both incipient and manifest disease, and enabled visits to children's homes by the school nurse to report on conditions and monitor progress.

In parallel with the intensive surveillance of the body of the infant during the early twentieth century, medicine also turned to focus on the unformed mind of the child. As with physical development, psychological growth was construed as inherently contingent and precariously normal. The initial solution was for psychological well-being to be monitored and its abnormal forms identified. The nervous child, the delicate child, the eneuretic child, the neuropathic child, the maladjusted child, the difficult child, the neurotic child, the over-sensitive child, the unstable child and the solitary child, all emerged as a new way of seeing a potentially hazardous normal childhood. The contemporary work of Freud that located adult psychopathology in early childhood experience epitomized the logic of a vision that saw the child as psychological father of the Man.

The idea of the child as someone characterized by growth and development meant that the identification of abnormality was not as clear-cut as under the clinical regime that separated physiology from pathology. A minor perturbation from the normal, in time, could become a full-blown abnormality. Medicine therefore had to look for those subtle suggestions of variation from the normal that might progress to pathology. Clinical examination of children tried not only to identify disease but also the many 'correctable' defects that might indicate the early stages of disease (Collins 1934: 321). Sometimes special techniques had to be devised to enable these defects to be distinguished from background variability. Placing the body under stress was one way, particularly during physical exercise:

> In order to detect physical deficiency, observation should be carried out during the course of physical performance. A child might be found easily fatigued, sluggish in movement, easily out of breath, clumsy ... symptoms in no way associated with pathological change (but) fine departures from the normal which would be undiscoverable in the course of ordinary clinical examination.
>
> (Holt 1928: 399)

The identification of 'early' abnormality pushed back the range of what might be considered normal; equally, the ubiquity of these incipient signs of disease meant that they became themselves part of the normal. In other words, the central distinction of clinical medicine, that between the normal and the abnormal, began to dissolve.

If there is one image that captures the essence of this reconstruction of the space of evaluation surrounding the child in the early decades of the twentieth century, it might well be the height and weight growth chart. Such charts contained a series of gently curving lines, each one representing the growth trajectory of a population of children. Each line marked the 'normal' experience of a child who started his or her development at the beginning of the line. Thus, every child could be assigned a place on the chart and, with successive plots, given a personal trajectory. The individual trajectory, however, only existed in a context of general population trajectories: the child was unique yet uniqueness could only be read from a composition that summed the unique features of all children. A test of normal growth assumed the possibility of abnormal growth: yet how, from knowledge of other children's growth, could the boundaries of normality be identified? When was a single point on the growth and weight chart, to which the sick child was reduced, to be interpreted as abnormal? Abnormality was a relative phenomenon. A child was abnormal with reference to other children, and even then only by degrees. Growth charts distributed the body of the child in a field delineated not by the absolute (and binary) categories of physiology and pathology, but by the characteristics of the 'normal' population.

These various observations of the child's body, of its achievements and its behaviour, placed it in a pluri-potential space far removed from the strict binary classification of the nineteenth century. Echoing the comparative logic of interpersonal hygiene the referent for any dimension of childhood was no longer nature but other children: how did this child compare with others in terms of height, in terms of mental development, in terms of examination results, in terms of athletic prowess, and so on. Each dimension was a scale with multiple points to which every child, at least in principle, could be assigned. Moreover, because there were so many such scales for the many attributes and characteristics of childhood a multi-dimensional space was opened up in which the child could be located. Just as the two dimensional growth-weight chart mapped early physical development against time, so the multiple axes of the new space brought all the significant characteristics of childhood under the monitoring eye of medicine.

By the early 1950s, the process of dimensionalizing childhood had reached a stage when a summative assessment could be offered.

In *The normal child* (Illingworth 1953) (fittingly republished every few years thereafter and translated into many languages), Illingworth provided a definitive statement of the new identity afforded every child: different but normal, diseased but normal, abnormal but normal. Whatever the extreme of a child on any one dimension, it was still a scale relating the child with others, and it was still only one dimension while the great web of 'normal' space lay all around.

Following the successful application of the new techniques to children, the way was open to draw adults fully into the multi-faceted space of identity. Adults, however, were not under the same sort of regular surveillance as children; it was realized that a regular health examination by doctors for example, was impractical in terms of resources and difficult in terms of technique (Britten 1931). Even so, the potential was now apparent: adults too could be placed in a normalizing space that would enable the identification and treatment of incipient disease.

> We see it in the proposal, not yet in general operation, that doctors should examine their patients, sick or well, periodically ... (for) the unsuspected beginnings of grave disease.
>
> (Currie 1938: 3)

Mapping population dimensions

There were two main techniques for applying the normalizing or 'dimensionalizing' process to the whole population: one was screening, the other was the socio-medical survey. The initial rationale for screening did imply a binary classification as normal was separated from abnormal findings but as with childhood, the identification of the early signs and symptoms of pathology translated formerly normal states into abnormal ones, and at the same time, given the frequency of such findings, brought about a re-definition of what was normal. For example, in the proto-screening regime of the inter-war Pioneer Health Centre in Peckham, south London, patients registered with the Centre were offered an annual consultation in which they were given screening for early or manifest disease (Pearse and Crocker 1943). Remarkably, only 7 per cent of those attending the Centre were found to be truly healthy, the rest had some abnormality or other of structure or function. After the war, with the extension of more formalized screening programmes – individual, population, multi-phasic, or opportunistic – the twin processes of 'abnormalization' and 'normalization' of the population continued in parallel. Sometimes a screening programme was established to sift through a population in search of a pathological lesion; other times it was to look for a pre-cursor of illness, a proto-lesion that might later transform itself into a full-blown pathology (Schenthal 1960).

While screening began to collapse the established distinction between physiology and pathology, the socio-medical survey both reinforced the idea of non-binary classification and extended the dimensions along which such judgements could be exercised. Survey technology had been developed early in the twentieth century but its potential for application to the problem of health was only realized mid century. In Britain in the 1930s, Mass Observation began to recruit respondents to record 'ordinary' events and their response to them. In the US, the 9000 family periodic health canvas began its annual sweep to establish the range of illnesses reported by the population (Collins 1933). In addition, in Britain the war-time Social Survey, set up to assess public opinion, added questions about the prevalence of ill-health.

The results of the socio-medical survey threw into relief the distinction between the biomedical model's separation of health and disease, and the survey's continuous distribution of variables throughout the population. Within this new analytic frame, health was not construed as a nominal category in a binary classification but as a scale, as an axis, as a dimension: health was not the opposite of illness, but something that could be improved or damaged by small degrees. As Oleson reported in 1939, analysis of 10,000 'recent questions' addressed to the information service of the US Public Health Service showed that 'in general, these figures indicate a shift in emphasis from consideration of disease as entities to those involving hygiene, sanitation, facilities for medical care, climate, diet and the like' (Oleson 1939: 765).

The tactics of clinical practice that had appeared in the nineteenth century had been those of exile and enclosure. The lesion marked out and separated (usually in the hospital) those who were different in a great binary system of illness and health, and then processed the ill in an attempt to return them to the other side, the side of health. The new perception recognized that health did not exist in a strict opposition to illness, rather health and illness belonged to ordinal scales in which the healthy could become healthier, and health could co-exist with illness. The logic of this perspective, when fully developed, would mean there was nothing incongruous, for example, in having cancer yet believing oneself to be essentially healthy (Kagawa-Singer 1993).

Large American community studies of psychiatric morbidity such as the Midtown Manhattan and Sterling County studies identified upwards of 60 per cent of the population as being mentally ill with the remainder having many transient psychological disturbances (Srole *et al.* 1962; Leighton *et al.* 1963). Further research found that between 30 per cent and 90 per cent of so-called 'organic' complaints had a psychiatric component (Tredgold 1962). Medicine could no longer imagine that segregating the insane would hold mental illness separate from 'sanity'. Asylum walls therefore marked no meaningful line in a classification system that placed everyone on a scale of degrees of psychological health: in consequence, the asylum's doors could be opened (in the 'reforms' of the 1960s) and its dividing walls could be eradicated from psychiatric practice.

A century earlier the patient had been no more and no less than the body that enclosed the lesion. The survey, however, embraced everyone, and found that almost everyone experienced symptoms most days of their lives even though very few were taken to the doctor (Mechanic and Volkart 1960). The concept of the 'clinical iceberg' that described those under health care as only the tip of an enormous mass of morbidity in the community was first advanced in 1963 and was confirmed and reconfirmed in subsequent studies (Last 1963). The transition from person-status to patient-status had been marked by 'coming under the doctor'. Now, as this interface began to fade, it received increased scrutiny to understand better a seemingly arbitrary transition (Zola 1973). The difference between a person/patient with a symptom who chose to use health services and someone with a symptom who chose not to reflected more on their behaviour than their health or illness status. Everyone was normal yet no-one was truly healthy.

The contrivance of multi-factorial aetiology matched a system of health based on multiple partitioning in the post-war years (which in its turn furthered the conceptual development of the field of health behaviours). For example, as disease became increasingly located within a social space there was concurrent growth of psycho-social models of causation. Social class was construed as a major determinant of ill-health and patient behaviour and stress became implicated in the development of a broad range of diseases (as did its 'antidote' of social support) from the 1950s onwards (Selye 1956; Janis 1958). Not only did psycho-social factors play a direct aetiological role but, from the early 1960s, theories of labelling and stigma charged that the illness state could arise at times without any physical mediation (Goffman 1961; Szasz 1962).

While the survey could show the distribution of illness in a whole population, its value was constrained by the limited availability of data and techniques with which to measure, evaluate and compare gradations in health. The only comprehensive medical records, for example, were those on mortality that represented the logic of a binary system. The survey demanded alternative ways of measuring illness that would encompass nuances of variation from some community-based idea of the norm. Hence the development in the last two decades of the twentieth century of a plethora of health profile questionnaires, subjective health measures, and other survey instruments with which to identify the proto-illness and its sub-clinical manifestations, and the deployment of qualitative methodologies that might capture illness as an experience rather than as a lesion (Fitzpatrick *et al.* 1984).

The survey classified health, illness, and bodies on a continuum: there were no inherent distinctions between a point at one end and one at the other; their only differences were the spaces that separated them. When Faber published his *Nosology in internal medicine in* 1924, he acclaimed the nineteenth century essentialist model of disease classification in which each disease was a separate expression of an internal pathological lesion, each its own space in a classification table of absolute entities. By the 1950s, this form of understanding was under strong attack from a nominalist perspective that held that disease was only a name given to certain configuration of signs and symptoms. Disease was not an immutable phenomenon but an arbitrary classification point in a great space of health and illness expression. Within the multi-dimensional field of medical observation there were many intersecting lines of characteristics that could coalesce into a disease, or hint of disease, in that great universe of spaces cross-cutting spaces. Perhaps at its most simple, the shift can be illustrated with the celebrated debate between Platt and Pickering on whether blood pressure was bi-modally or continuously distributed in the population (Pickering 1962). On the one hand, there was a nineteenth century reading of separate diseases with their own 'personalities' expressing themselves on the canvas of the body. On the other hand, there was the new view that disease was a convention, a particular way of classifying a perturbation in the physical and social space that enveloped identity.

When illness had been localized to a specific point inside the body of an individual patient the principal clinical task was the mapping of the three dimensional depths of the body using the techniques of physical examination. When illness moved out of its corporal confines and became distributed in the gap between bodies, in the interstices of the social, in the space that was to become known as the community, the task was to survey this social space. In the closing decades of the twentieth century this activity did not require the exclusive skills and organization of a clinical practice that focused on body interiors, rather the monitoring gaze of medicine needed to embrace a range of disciplines, many new, that sought to analyse a new configuration of disease and illness with a variety of novel techniques. The new dynamic of health meant that the whole population needed to come within the purview of a multi-dimensional surveillance apparatus.

At one level, the spread of multi-dimensionality throughout a population was a process of normalization in which everyone came under medical observation. In this guise, it elicited contemporary concerns about the dangers of extended 'social control' by medicine and the medicalization of everyday life (Zola 1972; Illich 1974). But these concerns were misplaced. In many ways, the term normalization, with its connotations of uniformity and control, was a misnomer. Normalization was only the outward manifestation of an exploratory technique that moved along the proliferating planes of inner identity. To become normal was to occupy an indeterminate and anonymous position somewhere within a multi-dimensional space. Perhaps when dimensions were few the position could be reduced to a certain 'character' or 'type' but as the axes multiplied even this device failed to encapsulate the myriad facets of identity.

The normal Man was not a point in multi-dimensional space – even less a nominal position in a binary classification – but a space of potential. The process of medicalization was not therefore a case of medicine invading a pre-existing everyday life but the very process of creating the multi-dimensionality of that life. Identity was becoming multi-axial and medicine simply followed the vectors that traversed the new space. Identity could no longer be captured by a unitary role or fixed personality; identity was made up of the planes and trajectories of a complex and opening space. When the WHO came to formalize its definition of health in the early 1950s – a state of complete physical, mental and social well-being, not only the absence of disease – is it at all surprising that it chose a state that was multi-dimensional?

11
Becoming at Risk

As the multi-dimensional space of identity unfolded, medicine maintained its watchful eye over the new spaces – physical, psychological, social, pathological, normal, etc. – that presented themselves to surveillance. Perhaps this elaborated and multi-faceted framework of medical observation can be seen as an outgrowth of the nineteenth century clinical examination: as the object of clinical attention was transformed so were the principles and practice of medical work, both locked in a dialectic and creative process. Alternatively, the cumulative effect of these changes might be judged as so profound as to warrant labelling as a qualitatively different system of medicine quite divorced from its nineteenth century roots. In other words, in retrospect, these various changes can be synthesized into a new model of medicine that placed the process of surveillance rather than the identification of the pathological lesion at its core.

Surveillance medicine

In the nineteenth century, the physician had to infer from symptoms and signs the type of pathological lesion lying within the patient's body. In consequence, the patient's body as a three-dimensional object became the focus of medical attention. This 'conceptual body' of the space of illness was matched by the construction of a parallel 'material body' in which illness was located. The classical techniques of the clinical examination that allowed the volume of the human body to be mapped in everyday practice and the spread of the post-mortem or autopsy as a procedure to identify incontrovertibly the exact nature of the hidden lesion provided the essential bridge between the conceptual and material worlds. These bodies of illness and techniques of diagnosis represented the medical model of disease, a model centred on the existence of the intra-corporal pathological lesion and the subsequent search for its true nature.

The new late twentieth century medical model – that might be described as Surveillance Medicine – was marked by an extension of the medical eye from scrutiny of the strict anatomy of the body to a great sweep across multi-dimensional space. This widening of medical vision, together with the deployment of new exploratory techniques, was accompanied by a parallel shift in the conceptual organization of illness as the relationship between symptom, sign and illness were reconfigured. From a linkage based on surface and depth, all became components in a more general arrangement of predictive factors.

For nineteenth century medicine, a symptom or sign was produced by the lesion and consequently could be used to infer the existence and exact nature of the disease. Surveillance Medicine took these discrete elements of symptom and sign and subsumed them under a more general category of 'factor' that pointed to, though did not necessarily produce, some future illness. Such inherent contingency was expressed by the concept of risk. It was no longer the symptom or sign pointing tantalizingly at the hidden pathological truth of disease within the body, but the risk factor opening up a space of future illness potential.

Symptoms and signs, therefore, were only important for Surveillance Medicine to the extent that they could be re-read as risk factors. Equally, the illness in the form of the disease or lesion that had been the end-point of clinical inference under the old medicine was also deciphered as a risk factor in as much as one illness became a risk factor for another. Symptom, sign, investigation and disease thereby become conflated into an infinite chain of risks. A headache might be a risk factor for high blood pressure (hypertension), but high blood pressure was simply a risk factor for another illness (stroke). Moreover, whereas symptoms, signs and diseases were located in the body, the risk factor encompassed any state or event from which a probability of illness could be calculated. This meant that Surveillance Medicine also addressed an extra-corporal space –

often represented by the notion of 'lifestyle' – to identify the precursors of future illness. Lack of exercise and a high fat diet could be joined with angina, high blood cholesterol and diabetes as risk factors for heart disease. Symptoms, signs, illnesses, and health behaviours simply became indicators for yet other symptoms, signs, illnesses and health behaviours. Each illness of pathological medicine existed as the discrete resultant in the chain of clinical discovery: in Surveillance Medicine, each illness was simply a nodal point in a network of health status monitoring. The problem was less illness *per se* but the semi-pathological, pre-illness, at-risk state.

Under the old medicine, the symptom indicated the underlying lesion in a static relationship; to be sure, the 'silent' lesion could exist without evoking suspicion of its presence but eventually the symptomatic manifestations erupted into clinical consciousness. The risk factor, however, had no fixed or necessary relationship with future illness; it simply opened up a space of possibility. Moreover, the risk factor existed in a mobile relationship with other risks, appearing and disappearing, aggregating and disaggregating, crossing spaces within and without the corporal body.

Where clinical medicine had operated within the three-dimensional corporal volume of the sick patient, the risk factor network of Surveillance Medicine was read across an extra-corporal and temporal space. In part, the new space of illness was the community that enclosed the grid of interactions between people. This multi-faceted population space encompassed the physical gap between bodies that needed constant monitoring to guard against transmission of contagious diseases; but the space between bodies was also a psychosocial space. The emergence of this new multi-axial field accompanied the shift in medical perception from the binary problem of health/disease to the generalized population problems of the symptom/illness iceberg. It was characterized by the crystallization of individual attitudes, beliefs, cognitions and behaviours, of limits to self-efficacy, of ecological concerns, and of aspects of lifestyle that became such a preoccupation of progressive health care rhetoric.

A further important feature of the new population space of illness was its emphasis on a temporal axis. Nineteenth century medicine contained temporal elements but relied essentially on a cross-sectional nosographic technique: patients had to be classified according to the nature of their internal lesion so that appropriate therapy could be introduced. Diseases, of course, had antecedent causes and they had resulting consequences, but these aspects of illness were analysed from the point of the present: what caused the lesion that presented (better to guide therapy) and what was the future prognosis for the patient? The new medical discourse of the late twentieth century opened up this static model so as to place illness in a wider temporal context. This analysis was particularly evident in the twentieth century fascination with the problem of development, especially in its relation to children (and later 'ageing'), but temporal concerns can also clearly be seen in the mid twentieth century invention of the category of 'chronic illness' as a major medical problem.

It was not until 1927 that Index Medicus, a database of medical publications classified by author and subject, created the classification category of 'Diseases, chronic'. In 1947, an additional signpost was added: 'Chronic illness. See Diseases, chronic'. Finally in 1957 the full heading 'Chronic disease' made its appearance and in the following years accumulated a series of sub-headings – complications, economic, occurrence, psychology, therapy, rehabilitation, etc. Corroborative support can be found in 1955 with the beginning of publication of the *Journal of Chronic Disease*. Its opening announcement identified a world of illness and attendant medical techniques, which, it argued, had not existed ten years previously. Chronic illness used to be a 'hopeless affair' but not now; it used to be a problem of old age, but now was seen to affect all age groups; the physician used to be the single most important factor in its treatment, now it was the patient. Later editorials extolled the virtues of prevention and multiple screening, they called for a shift in focus from hospital to community and they demanded closer scrutiny of personal and social consequences for the individual patient. Chronic illness became an exemplar of the sort of temporal problem embedded within a regime of Surveillance Medicine.

The temporal space opened up by Surveillance Medicine provided the matrix in which risk factors could materialize. Risk factors, above all else, were pointers to a potential, yet unformed, eventuality. For example, the abnormal cells discovered in cervical cytology screening or high cholesterol levels in a blood sample in themselves did not signify the existence of disease, but only indicated its future possibility. The techniques of Surveillance Medicine – screening, surveys, and public health campaigns – all addressed this problem in terms of searching for temporal regularities, offering anticipatory care, and attempting to transform the future by correcting physical aberrations and changing the health attitudes and health behaviours of the present.

Illness therefore came to inhabit a temporal space. Illness had a life history: from its subtle indicators of embryonic existence to a series of minor disturbances that indicated its early presence; from its 'pre-illness' forms to its first symptomatic appearance; from its overt clinical manifestations to the medical attempts to alter its natural history. The sub-division of prevention into primary, secondary and tertiary forms summarized the points at which medicine could intervene in the great new cycle of illness. The clinical techniques of the hospital had invested the three-dimensional body of the patient; surveillance analysed a four-dimensional space in which a temporal axis was joined to the living density of corporal volume and psycho-social worlds. Pathology in the past had been a physical lesion; in Surveillance Medicine, illness became a point of perpetual becoming.

The new public health

Just as Surveillance Medicine re-ordered the conceptual and practical field of clinical medicine, public health was redirected to address new dangers. Inter-personal hygiene and social medicine had succeeded in sanitizing the space between bodies but new threats appeared in the second half of the twentieth century that were not restricted to relationships with the Other. Risks now came from everywhere, but nowhere more important than those hazards that arose as a by-product of human interaction. The new space of danger was a sort of abstracted outcome of collective otherness:

> Pollution of the air and water by industrial waste, sub-standard housing and exposure to new chemicals of unknown toxicity are all parts of the modern environment which contribute to the major health problems of the United States.
>
> (Dearing 1953: 1149)

In 1962, Carson described an apocalyptic 'silent spring' in which nature was finally silenced by Man's pollution of the environment.

Poisoned water, fouled air, soil full of pesticides and insecticides found their way into the food chain and threatened all of nature's creation; substances manufactured ostensibly to control nature were beginning to destroy the very environment that Man now depended on. Carson's plea was the beginning of a deluge of environmental concerns that addressed the threat from the destruction of plant and animal species that were part of nature's bounty. In addition, they identified the harm that Man might cause himself by interfering with his mutual interdependence with the natural world.

Carson's alarmist concerns found immediate resonance in a new variant of public health that emerged in the 1960s alongside the new environmentalism (Ashton and Seymour 1988; Martin and McQueen 1989; Draper 1991). Under this new public health, the danger arose not from nature as it did under sanitary science or from other individual bodies as under personal hygiene and social medicine, but from the interactions of those other bodies with nature. In an inversion of the nineteenth century focus on maintaining a separation of corporal and non-corporal space, the new public health was not concerned with the intrusion of 'nature' into bodies, but with the incursion of the activities of those bodies into nature. The new public health medicine discovered that the by-products of economic and social activity could be dangerous and committed itself to maintain the purity of the natural environment. The dangers were everywhere and unavoidable: noxious gases from car exhausts; chemicals from aerosols attacking the ozone layer; acid rain from industry in the water and pollution in the soil; radiation from power stations, X-rays and work environments; electromagnetic fields from power cables, additives and harmful processing in food; 'unnatural' animal husbandry, genetic manipulation of 'natural' foods; everywhere unseen dangers.

The characteristic of each of these threats was that they represented a contamination of nature by human interaction/production. Nature was not the source of danger; indeed 'nature' was the wounded party. Moreover, whereas previous regimes of hygiene could fix on the hazard to minimize its threat and police a line or linear space of separation, the new dangers presented a major challenge to the maintenance of hygiene. Danger was everywhere: outside the body, inside the body, within the relationship of bodies, outside the relationship of bodies. Perhaps the contemporary epidemic of AIDS – and its accompanying panics – illustrated the breadth of the new threat. A virus within the body; a danger from exchange with other bodies, and, in addition, as the new public health emphasized, a wider context of social activities involving the socializing patterns or culture of gay men and the complex interactions implied by needle sharing and blood transfusion.

How best, then, to meet this pervasive challenge? There was no boundary to police, no *cordon sanitaire* or even permeable space to guard. A sanitary regime would not prevent solar rays from penetrating the body. No simple strategy of inter-personal hygiene could arrest or control the intake of food additives. The solution – and the problem – was behaviour, both individual and collective: behaviour as in political activity to promote 'environmentally-friendly' policies; behaviour to achieve a healthy 'lifestyle'; behaviour to be on guard against dangers from both within and without the self. This constant vigilance to dangers from everywhere involved a health promotion strategy with a 'green' focus, with an ecological awareness, that celebrated the space of nature not simply as the non-corporal but as a domain into which self reached out. Organic products, natural processes and ecological balances became the watchwords of the new hygiene together with perpetual alertness to the hidden threats that lay all around. Such individual and collective watchfulness might be described as a form of 'political awareness' or reflexivity. The new public health was concerned with generating, monitoring and maintaining this communal vigilance and ensuring that the subjective, reflexive body – or in terms of the contemporary slogan, the whole-person – was fully politicized.

Risks tracked along the planes of identity of the whole-person. Each observational tactic of Surveillance Medicine not only 'protected' health but also at the same time brought into clearer focus the axes of those once hazy dimensions of Man. The conceptual space of illness mapped onto the shifting morphology of identity just as surely as the nineteenth century medical model had reflected and sustained the solid three-dimensional density of anatomical Man. Together illness and medicine had defined, nay, invented the new anatomy of identity.

What, then, was identity? What an irony that the great battle of the nineteenth century to assert the supremacy of the Darwinian account of creation over the Biblical tale should have simply replaced one story of a distant history and a universal Man with another. At least the Biblical story added the Fall from Grace so that Man could thereafter try to transform himself spiritually in a search for redemption; but Darwin imprisoned Man in a three-dimensional body ruled by the dead hand of genetic forces of past generations that circumscribed his identity and potential. To be sure, ideas about genetics also found their way into Surveillance Medicine but these were newly expressed in the form of risks and conjoined with other risks from worlds unimagined by the great nineteenth century custodian of Man's destiny. What an irony that just as a new reflexive self materialized, a reinforced discourse on Man's biological roots emerged to stress further the universality of identity. From the 1970s 're'emergence of medical ethics that stressed transcendental morality (Beauchamp and Childress 1979), through the 're'visions of the New Darwinists in biology who attempted to reduce the new psycho-social components of identity to a genetic basis (Dawkin 1976), to the 're' surgence of a those in the social sciences advocating a closer alliance with Man's biological roots (Benton 1993), the new re-writing of the Darwinian creation story served to unify the multiple components of identity into that elusive whole-person. At the same time, it reasserted Man's narrative as a period spanning tens of thousands of years, rather than the brief moment of creation that was Man's all too recent history.

12
Death of the Old Hospital

The major changes in the medical model that had occurred by the late twentieth century found immediate expression in various shifts in clinical perception and practice, from the intimate encounter with patients to the wider strategies of health surveillance. In addition, the more salient organization of medical care also came to reflect the new medical framework. In particular, the nineteenth century centrepiece of clinical practice, the hospital, could hardly escape the reverberations of the collapse of the conceptual infrastructure that had for so long sustained it.

Hospitals had provided the 'neutral' space in which the indicators of the underlying pathological lesion might best be identified without contamination by the sort of extraneous 'noise' that was to be found in the patient's own home. Hospitals were therefore designed for the examination of bodies. Hospitals neatly separated bodies in individual beds where they were observed, monitored and examined; dossiers in the form of clinical notes were prepared on each body and regularly updated; clinical inference, investigations, visualization techniques and surgery allowed the unknown interiors of the body to be revealed to the medical eye; and in the hospital's post-mortem room the body was dissected to reveal its inner secrets.

Hospitals spread rapidly during the nineteenth and early twentieth centuries. In England and Wales, for example, there were 904 hospitals in 1861; by 1911 this had risen to 2,187, and by 1938 to 3,137 (Pinker 1966). This considerable expansion in hospital numbers was accompanied by an increase in size with the result that bed numbers rose at an even faster rate than the number of hospitals. In 1861, there were about 65,000 beds and by 1938 this had increased four-fold to 263,103.

The inter-war growth in bed numbers continued after World War II increasing to reach a peak of nearly half a million in Britain in 1960. Then the expansion stopped and a reversal began: there had been 455,138 beds in England in 1959, in 1980 there were 363,395, by 1990 255,054 (Department of Health 1992), and by 1999 only 186,000, a decline of over 50 per cent in 40 years.

The number of patients treated in hospital showed a different picture. In the mid nineteenth century average length of stay for a patient was 36 days in England and Wales and this had halved to 18 days by 1938 (Pinker 1968). Taken together with the greater numbers of beds that were available, this meant that an increasing proportion of the population was passing through the hospital. This rising 'throughput' more than compensated for the decline in bed numbers after 1960. Reduction of length of stay to 9.4 days in 1979 and to 6.4 days in 1989 (Department of Health 1991) ensured that overall there were increasing rates of 'hospitalization' throughout most of the twentieth century, a pattern reflected in other countries.

The broad picture, then, was one of decline in the numbers of hospitals and hospital beds in the closing decades of the century – but also increased rates of hospitalization. In part, this was a picture of greater efficiency as more patients were treated with fewer 'fixed' hospital resources, but such an explanation masks a more significant change, namely a fundamental transformation of the hospital as an institution from the early 1960s onwards together with a commensurate change in the meaning of hospitalization.

For over a century beds lay at the heart of a system of medical management based on clinical observation of patients' bodies by both medical and nursing professions. Beds provided the critical therapeutic space in which the art and science of medicine could be practiced – and the twentieth century witnessed a vast increase in the skills and technologies that were applied to the hospitalized patient. Beds allowed every individual patient to have a recognized place in the institution of the hospital, an address in the form of a ward name and a unique bed number. The rest of the hospital was built to service these beds and the individualized patients that they cradled.

Safety and rest

Following Nightingale's nineteenth century sanitary reforms, the hospital began to become a place of good hygiene in which good environmental conditions could be provided for the sick. The aim was to ensure that bodies and dirt were kept apart. Hospitals were therefore pre-eminent places for body construction: the hospital enabled the body to be mapped by clinical medicine, dissected in the post-mortem room, and separated from nature by strict sanitary rules. The hospital and its beds therefore played a central role in the invention of anatomical Man.

At the end of the nineteenth century a new component was added to the regime of safety that pervaded the hospital bed when it was recognized that not only did the hospital protect the patient by removing him or her from insanitary conditions, but it also protected the population by removing potentially dangerous and contagious patients from the midst of others. This new role for hospital care reflected the emerging contemporary concerns with inter-personal space and hygiene, of preventing bodies from contaminating one another. To this end, the hospital ideal was to establish a separated place, removed from everyday life, in which the dangers that the sick posed did not threaten the public but were contained and managed within the confines of the hygienic and ordered space of wards. Indeed this important role for the hospital was prefigured in the growth of specialized fever, infectious disease and 'isolation' hospitals that applied the great hygienic principle of quarantine to those patients who might be a danger to others.

The hospital removed the potentially dangerous patient from the home and population and applied more locally the principle of keeping bodies apart by using a strategy of separation within its internal organization. Just as the hospital isolated dangerous patients from the population, so the ward separated them from the rest of the hospital and further sub-division of the ward contained any hazard to the bed itself. In its turn, the bed could be subjected to a system of micro-quarantine through techniques such as 'barrier nursing' and control over the flow of visitors. The hospital could claim therefore to be twice over a place of safety in that it was both a sanitary space that separated the patient from potential dangers in the home and environment, and the hygienic device that separated dangerous patients from vulnerable others.

To the virtue of safety could be added a second core principle: that hospital beds were therapeutic. Certainly increasingly sophisticated medical treatments were administered to the hospital patient but in addition, the bed itself was held to have a therapeutic effect that augmented anything that medicine might offer. This therapeutic effect was well summarized and justified in Hilton's nineteenth century classic text *Rest and pain*.

Hilton, an eminent surgeon, first published a series of lectures 'On the influence of mechanical and physiological rest in the treatment of accidents and surgical diseases, and the diagnostic value of pain' in 1863. The book was reissued at regular intervals under a number of different editors but Hilton's *Rest and pain* remained an essentially unchanged classic right through to its final (sixth) edition in 1950. The fundamental thesis of the text was a 'most exalted admiration of Nature's powers of repair' (Hilton 1892: 3). Hilton provided many examples of how Nature ensured rest for damaged tissues so that self-repair could begin; for example, the spasm in the muscles surrounding an inflamed knee joint was evidence for Hilton that Nature was using mechanical means to enforce rest on the damaged tissue. Hilton was also of the opinion that the mind, like the body, could benefit from this regimen. He quoted from a letter from the 'late Dr Hood of the Bethlehem Hospital':

> I am frequently applied to for the admission of lunatics into this hospital, whose insanity is caused by over mental work, anxiety, or exertion, and for whose cases nothing is required to restore the equilibrium but rest. Therapeutic measures are not necessary, all the mind seems to need is entire repose.
>
> (Hilton 1892: 9)

In effect, the bed was itself a therapeutic space. Patients were subjected to enforced bed rest in the ready knowledge that this regimen would enable Nature to effect her repair to damaged tissues and thereby aid the patient's recovery. Any formal medical intervention was simply additive to this underlying therapeutic agency. Thus, safety and therapy were together joined in an alliance that located the hospital bed at the centre of health care. Clinical practice developed in this privileged space and the numbers of hospitals and hospital beds expanded to exploit this virtuous conjunction.

The bed was no less than the basic productive machine of the anatomical factory of the hospital. Whereas the construction of the anatomical body can largely be described in terms of separating Man from nature, the hospital bed provided a technology that allowed the raw and dangerous energy of nature to be harnessed to the very process of fabrication. Like the force field of the electromagnetic chamber that allowed physicists to contain and study the primaeval forces of nature, so the bed employed its rituals of separation to establish the protected and rarefied space in which nature infused the healing energy that made whole the patient's body.

From safety to danger

In the middle of the twentieth century, the first hints of a threat to the putative safety of the hospital can be identified in texts on infectious diseases. For example, in the 1940 first edition of their book on infectious diseases, Harries and Mitman offered a traditional description of the control of infectious diseases in hospital – quarantine, barrier nursing, exclusion of visitors – in which the danger arose solely from the body of the patient. However, by the second edition some four years later, they acknowledged for the first time that the hospital itself could be a source of danger:

> Attendants, particularly in hospitals and institutions, are liable to play an important part in the transmission of disease.
>
> (Harries and Mitman 1944: 37)

The second edition also offered a reassessment of the place of isolation in the management of infectious disease. In the view of Harries and Mitman, isolation hospitals had failed, in part because the separation came too late, and in part because it separated patients from the world outside, but not from each other. The dangers now came, they argued, from cross-infection, a source of 'further illness and additional perils', especially from the symptomless carrier on the wards. It had been reported that 19 per cent of patients acquired a new infection while on the wards from other patients. Their solution was an even more isolationist regime within the hospital involving mechanisms such as novel architecture, swing doors without knobs, and separate utensils for each clinical 'cell'.

What was new in this analysis was not the threat of direct patient-to-patient infection – this had been a problem devised and addressed by inter-personal hygiene much earlier in the century. The new dangers came from the hospital itself as cause or at least key intermediary in the process of transmission of disease. Texts were devoted exclusively to the problem; cross-infection as mediated by the hospital became a major blight on the reputation of the latter. Hospital care was beneficial except for 'one serious drawback – that the disease from which one is suffering may be transmitted to others' (Williams *et al.* 1960: 1). In the new search for the ways in which the hospital might be a threat to patients' well-being, suspicion fell particularly on staff and ward objects. Williams and his colleagues (1959) showed that whereas 38 per cent of patients carried the bacterium staphylococcus aureus in their noses, some 68 per cent of hospital staff did so (and 52 per cent carried strains resistant to penicillin). The problem of hospital blankets became a *cause célèbre* on which could be articulated further concerns with the dangers of cross-contamination of bodies. In 1957, Frisby complained that there were then only three criteria for washing a blanket: the patient had died, an obviously infectious patient had last used it, or it looked dirty. But what of the hidden infection from ordinary patients? The blanket could come into contact with one patient and then another passing on any dangerous organisms.

> The hospital staphylococcus must be attacked before it reaches the patient and one of the points of attack must be the blanket.
>
> (Frisby 1957: 508)

Barnard (1952) had already pointed out that freshly laundered blankets contained dangerous bacteria and various experiments in better washing were underway (Thomas *et al.* 1958). In 1960, Anderson and his colleagues artificially infected two beds on an eight bed renovated (and empty) ward; after bed-making the infection was found to have travelled to all the other beds (Anderson *et al.* 1960). If bacteria could take to the air with such ease then they could lie dormant on floors, and they could float into 'clean' areas such as operating theatres. A burgeoning literature discovered ubiquitous dangers, from tetanus spores in operating theatres (Lowbury and Lilley 1958) to faecal contaminants in trouser turn-ups. In 1959 the Standing Medical Advisory Committee of the British National Health Service Central Health Services Council released a report, *Staphylococcal Infections in Hospitals* (Standing Medical Advisory Committee of the Central Health Services Council 1959), identifying a situation that an editorial in the *British Medical Journal* claimed was 'now recognised to be a very serious problem' (*British Medical Journal* editorial 1959: 218).

At one level, the danger of cross-infection was a technical problem, one engaged in by laboratory-based bacteriologists who scurried round the hospital, identifying organisms, disinfecting blankets, and debating the relative value of brushing and vacuuming floors, but it was the very existence of these minor technical concerns that marked the beginnings of the crisis of confidence in hospital safety. Over a period of about 20 years, and largely predating the actual decline in numbers of hospitals and beds, the hospital itself came to be seen as a source of infective material. Whereas previously there had been a medical rhetoric proclaiming the ideal of a hygienic hospital dealing with the dangerous patient, it was now the patient at risk from the dangerous hospital. Moreover, despite the contemporary introduction of antibiotics, the solution to the problem was seen to lie in a more penetrating analysis of the whole hospital:

> We realise now that to achieve this end we must look again at all aspects of the hospital, from the planning of the building at one end to the finest details of our sterilisation methods, to the techniques of the operating suites and wards, and to the education of all who work in a hospital.
>
> (Williams and Shooter 1963: 6)

From therapy to harm

In 1942 Atkins advised that there were no fixed rules for bed rest after a surgical operation but that even younger patients, who were the most anxious to get up, found themselves very weak after they had done so (Atkins 1942). Accordingly, he recommended between seven and ten days bed rest after 'interval' (elective) operations and 2–3 weeks after more severe operations. At the end of this period, the patient could gradually be mobilized: first, a patient could get out of bed while it was made, the following day up for tea, slowly increasing the time until discharge. Even at home, the convalescing patient should aim to have 'breakfast in bed and if possible a holiday in equal duration to hospital stay'. Problems that later writers would ascribe to prolonged bed rest were largely absent from Atkins' account and even when they received mention – such as the management of bed sores – their existence was in no way linked to lying in a bed.

A decade later in the fourth edition of his text, published in 1952, Atkins revised his advice on extensive bed rest. Instead, he commended early ambulation:

> Early rising diminishes the incidence of most vascular and pulmonary complications and speeds up recovery and reduces the period of convalescence.
>
> (Atkins 1952: 22–3)

In other words, between about 1940 and 1950, the logic of Hilton's *Rest and pain* was reversed. Not only did rest bring about harm rather than healing, but less post-operative rest reduced the later need for convalescence. The instructions were now that:

> After most operations a patient can get out of bed on the following day while his bed is being made, and thereafter for increasing periods.
>
> (Atkins 1952: 23)

It was not only surgeons, however, who identified the potential harms of bed rest. In 1947, a physician, Asher, described 'major hazards of the bed' affecting almost all body systems:

> The end result can be a comatose, vegetable existence in which, like a useless but carefully tended plant, the patient lies permanently in tranquil torpidity.
>
> (Asher 1947: 967)

In short, to the dangers of cross-infection could be added the not inconsiderable hazards of bed rest. These were considerable in their range and extent and included:

> Deterioration in morale, discomfort of constipation, abdominal distension and urinary infections, the liability to peripheral venous thrombosis, the lack of muscular tone and the development of arthritic adhesions.
>
> (Gilchrist 1960: 215–216)

Here was the irony of an institution whose existence was justified on the benefits it conferred – or, at the very least, the dangers it held at bay – finding that it could be a threat to the physical welfare of its patients. The hospital was no longer a place of safety from infections; bed rest was no longer the therapeutic process it had once seemed. Dangers seemed to lurk in the very fabric and routines of hospital life. Even staff posed threats as was demonstrated by the conviction for manslaughter of an anaesthetist addicted to inhaled anaesthetics: in the words of a *British Medical Journal* editorial, there was a need for a 'machinery to assist in preventing harm to patients from physical or mental disability, including addiction of hospital medical or dental staff' (*British Medical Journal* editorial 1960: 1797).

The routines of the hospital also threatened patients' psychological health and a series of reports and publications during the 1950s and early 1960s proposed measures to humanize the hospital, to counteract its alienating customs. For example, a report from the British Ministry of Health (1961a), *The Pattern of the In-patients Day*, was critical of the policy of waking patients at an early hour to ensure that they were washed, fed and ready for inspection by 9.00am. Contemporary concerns with the psychological dangers of hospitals were reflected in reports on noise in hospital and its stressful effects (King's Fund 1958), on children in hospitals and the unsuitability of the hospital environment (mostly uninviting, with often severe discipline and, of course, parental separation) (Ministry of Health 1959), and on 'human relations' in obstetrics (arguing the case for improvements in the management of noise, visiting hours and 'communications') (Ministry of Health 1961b).

In 1951, Asher described a new syndrome, Munchausen, in the pages of the *Lancet* (Asher 1951). The main symptom of this unusual disease was the pursuit of hospital treatment; Munchausen patients would feign illnesses so that they could be admitted and treated, often with surgery. It was, in Blackwell's words, a sort of 'hospital addiction' (Blackwell 1962). In both image and substance, the hospital had become a dangerous object.

For two hundred years, its proponents had advanced the hospital bed as the solution to illness but now, within a few years around the middle of the twentieth century, its core logic began to collapse. The two vital principles of safety and bed rest had been undermined as a search for dangers began to take hold. These searches were neither political nor economic in focus; it was not a covert agenda to destroy the hospital that underpinned the studies of hospital floor cleaning and measurement of ward noise levels, but the net effect was to build a wide-ranging technical critique of the hospital and the hitherto hidden costs of its care.

Hospitals and bodies

The hospital had long been at the centre of health care provision and the medical management of illness. Established as part of the late eighteenth century and early nineteenth century medical revolution that localized illness to the three-dimensional confines of the human body, the hospital acted as one of the key operational sites of corporal transformation. Patients were confined to beds that allowed the clinical examinations of medicine and the regular observations of nursing to act in an unimpeded field of visibility. But more: the application of sanitary regimes of public health – that great nineteenth century process of delimiting body boundaries – could reach a state of near perfection in the totally controlled environment of the hospital. Sanitary science outside the hospital would always express itself in broad-brush strokes, in vast sewage systems driven through teeming Victorian cities, in national bureaucratic hierarchies of inspection, in rhetorical pronouncements on the need for cleanliness. But within the hospital sanitary science could be applied in meticulous detail: the clean starched cap and apron of the nurse, the white coat of the doctor, the hand scrubbing, the ablutions, the removal of flowers at night to keep the ward air pure, the constant smell of disinfectant, the polished floors, the never-ending cleaning and washing. Outside the hospital walls, sanitary science often failed in its mission to bring everyone into a dirt-free world; inside the hospital walls was the possibility of a total sanitary regime from which no one could escape.

Whereas the force of sanitary science outside the hospital weakened in the early twentieth century, in its sanitized interior it continued, embedded in disciplined routines and practices. True, the new public health regime of inter-personal hygiene did make an impression in the even greater attention to contagious dangers, but in its self-imposed isolation the hospital remained a repository of archaic practices, a museum of techniques that since the mid nineteenth century had been used to forge the body of Man.

It was still the model of sanitary science, a dream of perfect cleanliness, that informed hospital architecture and routines right up until the middle of the twentieth century. It was still the passive and inert body of the patient lying in the hospital bed that underpinned clinical care, while outside the hospital walls movement and exercise were being practiced and promoted. Therefore, the rapid crisis of confidence in the effects of the hospital's sanitary regime and the crumbling of its *raison d'être* through the inversion of the relationship between bed rest and danger is not wholly surprising. The hospital's great work of manufacturing corporal space was largely completed; the cutting edge of medicine had moved on to social spaces, to revealing subjectivity, to producing reflexive identities. A new and different patient had emerged from the chrysalis of the hospital bed.

Reinventing the hospital

There are strong parallels between the re-assessment of the general hospital bed regime and the contemporary crisis in psychiatric hospitals. A policy of closing the old asylums was instituted in the 1960s, but the major difference was the location of danger In the psychiatric hospital it was not the bed – indeed the bed was, by and large, not a place of therapy nor did it become, in any reversal, a site of danger. In the main, psychiatric patients led relatively free lives within the confines of the asylum, retiring to a bed only at night. The asylum therefore represented a basic quarantine system, an old principle applied to the segregation of the mad to protect the sane. Danger came neither from 'dirt' nor from other bodies.

During the early twentieth century, madness began to be overtaken by the new 'inter-personal' problems of the neuroses that showed increasingly problems of coping presented the major threat to mental health. Where a nineteenth century psychiatric text would only deal with madness and its variants, by the middle of the twentieth century, the neuroses had displaced insanity from centre-stage. The neuroses, mainly anxiety and depression, were everywhere, requiring a constant mental hygiene in the home, in the workplace, and in the community, and they evinced the deployment of Surveillance Medicine to fight depression even before it appeared as a clinical entity. For its part, the asylum remained co-terminous with the space of madness, and when the space of madness collapsed so the asylum became anachronistic.

The problem with the asylum was not therefore its bed regime but the incarceration and isolation of the insane. Without the health problem of madness there was no means for the asylum to reform itself; accordingly the whole system had to go, asylum and all. In contrast, the crisis in the general hospital could be largely focused on the problem of the hospital bed and its surrounding practices so that the building could remain if its purposes were redefined.

The result of this 'contained' danger was that the hospital could survive in a form so long as the problem of the bed was addressed. Perhaps it would be easier to analyse these survival tactics if the hospital had changed its name so that the new could be distinguished from the old. If there were the Old Hospital and the New, it would be easier to see the points of transition as one replaced the other and to recognize the reformulation of meaning and experience implied by the term 'hospitalization'. Nevertheless, even with the constraints of an unchanging descriptor, transformation can be identified in the novel forms of 'hospital' that emerged.

The solution to psychiatric incarceration was the destruction of the surrounding walls to break the quarantine barrier that for centuries had kept madness contained; but the general hospital walls did not represent the same sort of problem. Nevertheless, in its new state of transition the boundaries of the hospital began to become more permeable. Walls, gates, porters and reception desks that had marked the interface between the hospital and non-hospital outer world became less rigid; hospital functions such as laundry and catering increasingly were provided by outside agencies; formalized referral mechanisms that had governed the transfer of patients from generalist to specialist began to be superseded by 'direct access' and 'outreach clinics'; a succession of reviews of the Casualty Department or Emergency Room took place – that intermediate space across which many patients passed in their transition from ill person to hospital patient; and a new emphasis on 'emergency medicine' made the hospital a nodal point in a system of triage.

The actual hazards of the bed were mainly addressed by restricting access, by minimizing time spent prostrate, by making the bed a place other than the traditional centre-piece of the hospital. Restricting access was a solution for the management of older patients who were now seen to have carried high risks in terms of their long-term hospital care: much better if they could be mobilized and discharged, perhaps to a Day Centre. Five-day wards were reported from the early 1970s that allowed patients home at weekends. Programmed investigation units were set up from the mid 1970s to enable patients to be admitted to a bed, usually for a brief period, so that they might undergo specific hospital-based investigations. Specialized admission wards in the form of either emergency admission wards or pre-operative planned admission units were tried in the 1970s. Of greater long-term importance, however, was the growth of day care in which the patient occupied a bed for a particular procedure but then returned home in the evening.

Another solution was to transfer the care package that surrounded the bed to the patient's home so that some of the benefits of the hospital could be preserved and the disbenefits eliminated. In her overview of the origins of the 'hospital-at-home' Clarke commented on 'the growing concern ... about the harm which may be caused by removal to hospital at times when we are particularly vulnerable to pressure from which home and family traditionally protect us' (Clarke 1984: 7). The idea spread into night nurse services, community hospitals and voluntary home care schemes.

Traditional health care statistics had been dominated by beds – separate figures were provided in Britain for 'available beds', 'staffed beds' and 'occupied beds' – but the emphasis gradually shifted to activity ('discharges and deaths') to reflect the numbers passing through hospitals rather than those staying in them. In 1977, the Department of Health began reporting the numbers of cases treated in the hospital and in 1991 changed the measurement currency to 'finished consultant (specialist) episodes'. Beds were becoming an irrelevance to measuring and understanding the work of the hospital.

Institutional decline

The Old Hospital died. From mid-century, a crisis of confidence in the hospital and its beds in the form of challenges to its claimed safety and therapeutic mission preceded a rapid decline in bed numbers. The result was that a New – and possibly transitory – Hospital emerged that addressed the central problem of the Old Hospital by emphasizing restricted bed usage and the development of more ambulatory services. At the same time, the disciplined space of organization and observation that had surrounded the bed began to dissolve as a series of studies and reports searched for different methods of hospital work (Joint Working Party 1967; Kogan *et al.* 1971).

This is not to say that there was doubt about the value of hospital-based therapies and investigations as these continued, albeit often administered in different ways, for example as illustrated by the rapid spread of day surgery. To that extent the post-war crisis does not seem to have had a technological basis in that the technical means of investigating and treating patients was maintained and developed throughout the period. Rather the problem seemed to be localized to the perceived increased dangers arising from keeping a patient in a bed or bringing them into hospital in the first place.

The decline of the Old Hospital was rapid. In the 1930s, Evans and Howard could foresee, with Messianic fervour, a 'conception of the British Empire as one vast hospital movement for the material redemption of mankind from those powers of darkness of hygienic ignorance and superstition!' (Evans and Howard 1930: 328). Indeed, they could cite Burdett's entry on 'Hospitals' in the Encyclopaedia Britannica: 'Why should we not have – on a carefully selected site well away from the towns, and adequately provided with every requisite demanded from the site of the most perfect modern hospital which the mind of man can conceive – a 'Hospital City'?' (cited in Evans and Howard 1930: 327–8). The model for this clinical utopia was the planned expansion of St. Bartholomew's Hospital located in the City of London. An institution affectionately known as Bart's, claiming to be one of the oldest hospitals in the world with a history going back nearly a thousand years, launched an appeal for a million pounds to prepare for its future. The vision involved a massive rebuilding programme to produce a 'city in a city', a hospital that was so large and so self-contained that it could offer patients and staff alike a total medical environment, a total medical experience. Sixty years later a decision was taken to close St. Bartholomew's Hospital. Despite appeals to its history and continuing ambitious plans for its future, it was proposed that it be 'merged' with a rival hospital several miles.

What happened to this 'city within a city'? What force could destroy a thousand years of history with such rapidity? What lies between the inter-war 'romance of the voluntary hospital' (to use Evans and Howard's term) and the emptying wards? These questions apply to Bart's, but also to hospitals and hospital beds in general that were closed at an ever-increasing rate during the final decades of the twentieth century.

In part, the demise of the Old Hospital lay in the crisis of confidence that overwhelmed its heroic mission. More fundamentally, however, it lay in the transformations in body and identity that made the Old Hospital a redundant institution. Just as the demise of the asylum can be linked to the disappearance of madness and the emergence of psychiatric disorder in the community, so the decline of the traditional hospital in the immediate post-war years can be seen as simply the architectural and institutional manifestation of a new perception of illness and of a new identity for the patient. The decline of the Old Hospital paralleled the rise of Surveillance Medicine; they were but two sides of the same coin. The great binary separation of the healthy and the ill that had sustained the hospital was replaced with a continuous distribution of health/illness, in which everyone was precariously healthy and everyone was precariously ill. From community surveys of the prevalence of ill-health to the new concerns with normality, the cognitive underpinnings of the Old Hospital collapsed. What role then for a bounded hospital? The question at once laid open the Old Hospital to the sharp arrows of those technical debates and anomalies that had seemed so very remote from the visionary city within a city.

13
Birth of Primary Care

The groundswell of crisis surrounding the hospital bed reached its crescendo in the early 1960s as the Old Hospital went into rapid decline. With its demise, the landscape of medical care provision was transformed. The hospital was reborn in a new form, a site of intervention less reliant on the bed and more on the type of investigation or treatment. More importantly, the New Hospital came into existence, in opposition, so to speak, to a new form of health care. Throughout the ascendancy of the Old Hospital, the rest of health care provision was a subservient and subsidiary activity: the body in the bed was the central axis of medical care and other forms of health care were merely adjuncts, preliminaries to the ceremony of bed-based clinical work. In contrast, the New Hospital belonged to a more equal division of health care that separated primary from secondary care.

An important part of the search for alternatives to hospital care was to reassess existing non-hospital health services and re-designate them as 'primary care' (so making the hospital part of a parallel domain of secondary care). The term 'primary care' had been used earlier in the century to describe the proposed feeder units in the constellation of the secondary care hospital (Dawson Report 1920) but in Britain, the term was revived in 1970 when the British Medical Association published a booklet on Primary Medical Care (British Medical Association Planning Unit Report 1970). The *British Medical Journal* reported some initial unhappiness with the new term (Brirtish Medical Journal editorial 1970) but by 1976 Hicks could use the words as the title of his overview of non-hospital services as it was by then 'generally accepted' (Hicks 1976).

This does not mean that the term was clearly defined; confusion was only to be expected if primary care was defined not in itself but in its relationship to the hospital. Primary care might or might not have included general practice, community health care, social services, the voluntary sector, self-help, and so on, but its defining characteristic remained that it was non-hospital. Pursuit of a strategy that gave preeminence to primary care was, above all else, a policy that challenged the hospital ascendancy. In Britain, a policy of encouraging primary care produced the wellknown 'Renaissance' of general practice; in the US, it involved support for 'family medicine'; world-wide, it took the form of the WHO's *Health for All* policy that stressed the central role of primary care.

General practice

The main element in the newly constituted field of primary care was general practice. Yet general practice was not simply recruited to help fill the empty spaces on the patchwork quilt of primary care: general practice was itself part of the transformative process. British general practice has written its own story of origins that begins in the early nineteenth century though its everyday activities probably changed very little until the 1950s. Then, a two-fold transformation occurred: not only did general practice realign itself in relationship to the hospital, colonizing the new space of primary care in the process, but it also began to reorganize its own internal spaces of clinical activity.

A flavour of the old world of British general practice can be glimpsed in a description of an inter-war practice. In his James Mackenzie Lecture of 1957 Hughes described the surgery (or office) of an 83-year-old doctor that he had joined as a young general practitioner (GP) in the mid 1930s (Hughes 1958). The setting, he surmised, had probably hardly changed in the 58 years that the old doctor had been there.

> The surgery was a room leading off from his smoke room; the walls were distempered in a dirty dark red; the floor was of bare boards and the room ill-lit by a small gas jet from his own plant. It contained a desk which was rarely used, half a dozen chairs, on one wall a dresser-like collection of shelves. ... There was no

> examination couch and no washbasin ... the old patients adored him and gladly waited for hours, half a day or longer if necessary, sitting on a stone bench around the pump in the yard, if the weather was fine, or on the chairs in the surgery if wet.
>
> (Hughes 1958: 5–6)

Hughes' inter-war surgery was a part of the domestic domain, an integral part of the old doctor's house: patients gained access through the GP's smoke room and without a couch or washbasin there was probably no means for a stranger to recognize that the ill-lit room was in fact a surgery. Clearly, this was not the hospital. There was minimal functional differentiation between the doctor's own house and the room in it that doubled as a surgery, and within the surgery space itself. To be sure, 'around the pump in the yard' seemed to have functioned as an informal waiting area but even this was dependent on the weather, as when it rained patients would sit on chairs in the same room in which the old GP would do his consulting. In effect, medicine outside the hospital was a domestic activity, hardly separable from GPs' and patients' homes; indeed, with the large frequency of home visits the patient's own domestic space was a contiguous part of general practice. Illness outside the hospital was therefore located in domestic spaces and domestic bodies. Yet the fault lines appearing in the hospital in the latter half of the twentieth century meant that general practice could not remain an indistinguishable part of the domestic.

The move away from domestic space involved a distancing from both the doctor's and patient's homes to establish general practice in a new community site. The first step in this movement away from the private worlds of doctor and patient can be illustrated by the description of a surgery that Handfield-Jones had created on appointment to a single-handed practice in 1954 (Handfield-Jones 1958). He had purchased a house that had two rooms at the back with a separate entrance. These two rooms became the consulting room and the waiting room:

> The waiting room is furnished with folding wooden chairs and a table for magazines. The walls are painted with white emulsion paint and the floor covered with light coloured linoleum. There are bright red plastic curtains with a floral design at the windows.
>
> (Handfield-Jones 1958: 206)

In the consulting room could be found the doctor's desk:

> placed to get the best light from the two windows. ... Behind the desk is the filing cabinet in which the record cards are kept; it can be reached from the chair. The couch lies along the wall. There is a space between the desk and the couch for the patient to undress, and a curtain on runners screens off this area from ceiling to floor ... the room is decorated in a restful shade of grey emulsion paint, the floor is of wooden blocks. Two big windows look out on to the garden and have gaily patterned curtains.
>
> (Handfield-Jones 1958: 209)

Handfield-Jones' surgery marked a transitional phase: it was still a part of the GP's house, yet by having a separate entrance for patients, it had become more of an appendage to domestic space than an integral component. Moreover, despite the clear domestic affinities in terms of decor and homeliness the two rooms that the patients used were specifically designated for medical use and did not appear to have had any alternative domestic functions. In addition, by the use of lobbies and doors marked 'Private' a strong physical barrier was erected between medical and domestic space. He separated off a specialized area for 'waiting' from the consulting room proper although the temporary seating in the form of folding wooden chairs signified its provisional status. Handfield-Jones was quite clear on the separation he had achieved: 'In these circumstances the practice does not intrude unduly into private life' (Handfield-Jones 1958: 211).

The surgery annexe also marked the appearance of new, though relatively weak, spatial sub-divisions. In the consulting room, patient notes were stored in a record cabinet within easy reach; patients were examined on a couch that was temporarily separated from the rest of the room by means of a curtain; and dressings and equipment for minor treatments were to be found in a cupboard alongside the drugs that the doctor would himself dispense. Certainly this was a medicalized space, its order and facilities mirrored the hospital in certain elements of its organization – couches, curtains, medicine cupboards, and the like would have given an appearance of clinical space – but it was still only a step away from the domestic space to which it was attached by a short path, a corridor, or a sign on a door.

Then, just as the hospital began its period of steep decline the space of a separated general practice – in a new invigorated form – began its consolidation. A conventional date to mark the transformation of general practice in Britain would be 1966, the year of the GP Charter, when 'health centres' became a part of the primary care landscape. Health centres or 'group practice premises' had been tracking the emerging crisis in the hospital and were already being built in prototype form in the 1950s. For example, Richard described an early yet typical health centre that had opened in 1955 (Richard 1962):

> The general practitioner unit consists of five suites of rooms, each suite comprising waiting, consulting and examination rooms. At the entrance to the corridor leading to the suites is a well-equipped dressings room with a nurse on duty. ... Each general practitioner's surgery is furnished with a desk, three chairs and all the usual diagnostic instruments ... the examination rooms are equipped with an examination couch, stool, table for instruments, and a wash hand basin. ... The clerical and administrative work is conducted from the small office staffed by a secretary and two clerks, all record cards are filed here.
>
> (Richard 1962: 257–8)

Such a configuration of space was rapidly to become the model for all future health centre or group practice premises. In the health centre, home and surgery were completely separated such that there was not even the possibility of patients crossing the boundary between medical and private space. Moreover, space was strongly differentiated. Every GP had his or her own waiting room; each activity, reception, nursing, examination, treatment, took place in a completely different room, if not building.

Whereas Handfield-Jones carried out all general practice activities himself, in the health centre those same activities were separated both physically and occupationally. The nurse, doctor, receptionist, pharmacist, etc. had separate quarters and separate functions. The health care team, whose growth paralleled the spatial reordering of general practice, in effect, had emerged to populate and administer the new medical space, intermediate between home and hospital. In the late 1959s there were less than twenty 'attached' paramedical staff in the whole of Britain but a decade later there were several thousand (Watson and Clarke 1972).

The boundaries and their markers that had mainly served to separate medical from domestic in Handfield-Jones' new post-war practice were now deployed to sub-divide and order the new space. The new health centre fragmented the patient and the illness into a series of modules:

> On entering the centre, the patient states his name and address, and the name of his doctor. He then proceeds to the appropriate waiting room and his record card is delivered by one of the office staff through a letter box of the doctor's consulting room.
>
> (Richards 1962: 258)

The patient would then be called into the consulting room to provide a history, sent into the examination room if an examination was judged necessary and directed to either the nurse in the dressing room or the pharmacist in the chemist's shop for treatment, or the administrative office again if another visit was called for.

Seemingly independent of this change in surgery arrangements, and yet exactly paralleling it, the GP also began to move away from the patient's own domestic space. For example, in the 1950s about one third of all GP patient contacts were represented by visits by the GP to the patient's home (Handfield-Jones 1959), by the mid 1960s it was down to about 20 per cent (Cartwright 1967), and by the 1980s closer to 10 per cent (Cartwright and Anderson 1981). In effect, with the post-war growth of health centres and separate practice premises, and with the decline in home visiting, a new space of medical work opened up midway between home and hospital.

A new location of illness

The spatial realignments of British general practice in the post-war years subjected illness to a new analysis, this focus involved the separation of illness from the domestic and its subsequent fragmentation. Illness was thereby reconstructed as a new phenomenon. The old general practice provided a service to individual familiar bodies: they sat chatting in the yard or on the folding wooden seats while they waited. Illness was located within each of these separate 'domestic' bodies. The new space of the health centre however only analyzed separate bodies to the extent that it compartmentalized them. The space it mapped onto was not primarily the physically discrete patient's body but a new target in the form of the patient's biography that was, in its turn, located in a wider context of 'the community'. Biography was elicited in the new enlightened forms of consultation that prioritized the patient's subjectivity but the idea of the community in which both biography and primary care co-existed was an even more recent invention.

The community was not a phenomenon that existed independently of the social perceptions that constructed it. The problem is partly one of disentangling the changing meaning of the word community in the context of its contemporary usage, but also, and especially for general practice, it is one of identifying when the word itself began to be used generally as a descriptive term. Certainly, in the 1950s, when Doctors Hughes, Handfield-Jones and Richard were describing their surgery buildings, the term community was not in general use. Handfield-Jones referred to his practice as covering 'nine villages in an area of two and a half miles radius' (Handfield-Jones 1959: 205) while Taylor in a contemporary classic study defined a key criterion of a GP as someone looking after 'people in a well- defined area' (Taylor 1954: 546). Even when the word community was used, it tended to refer to a specific 'domestic' space. Thus, for example, in their account of the Peckham Experiment, Pearse and Crocker stated that a community was 'a specific organ of the body of society and is formed of living and growing cells the homes of which it is comprised' (Pearse and Crocker 1943: 292). Indeed, the more common use of the terms 'family doctor' and 'domiciliary care' perhaps best reflected the dominant domestic orientation of British general practice in the immediate post-war years. This can be seen, for example, in the 1963 report of the Gillie Committee on 'The field of work of the family doctor' which seldom made reference to the community and when it did so seemed to be using the word simply as a synonym for society (Gillie Report 1963).

Yet by the time of the 1971 report on 'The organisation of group practice', the expression 'medical care in the community' was used with seeming confidence (*The Organisation of Group Practice 1971*). Indeed, by the late 1960s medicine in general had begun to recognize the community as a space of illness and a space of clinical practice: the word 'community' finally became a medical subject heading in Index Medicus in 1967 when the terms 'Community health services' and 'Community mental health services' were introduced to replace the old 'Public health'.

This transition in the space of illness was also clearly illustrated in two national surveys of British general practice. In the first, published in 1967, Cartwright devoted a chapter to 'Family and domiciliary care' (Cartwright 1967); in the second survey, published in 1981, the term domiciliary care had disappeared (to be replaced by 'home visits'); furthermore it was reported that younger doctors were less likely than older to value the notion of 'family care' and more likely to want greater emphasis on 'community care' (Cartwright and Anderson 1981). Even the US preference for the term 'family medicine' to describe primary health care came under challenge (Schwenk and Hughes 1983).

The notion of the community as an origin, location and place of therapy for illness was not exclusively a fabrication of the new general practice but in its particular realization as a space coterminous with the practice population it certainly was. The GP's list of patients was not an amorphous substrate in which flashes of illness periodically and randomly showed themselves within the private world of domestic space, but was itself the space of illness. 'Each population', noted Scott in his 1964 Mackenzie Lecture, 'is in effect a population at risk' (Scott 1965: 14). Illness no longer struck unexpectedly at an individual body but haunted an entire population. In a regime of sporadic disorganized illness, each patient was, within his or her domestic space, a separate incident; under a regime that viewed the community as the space of illness then illness became a calculable and summable risk. Surveys of case-load and morbidity in the 1950s marked the beginnings of this new perception (College of General Practitioners 1962); practice disease indices in the 1960s and the spread of age-sex registers in the 1970s signalled its general extension.

Temporalizing practice activity

In a 'Coronation Issue' of *The Practitioner* published in 1953, a photograph of a modern waiting room was accompanied by the caption: 'the whole scheme is bright, clean and radiates efficiency' (Practitioner 1953). The photograph might well be said to show a bright and clean room but how could a particular spatial arrangement of tables and chairs 'radiate efficiency'?

The problem was that the photograph could only capture in its two-dimensional form the three-dimensional room. What it could not portray was the reconstructed temporal space that pervaded the apparent solidity of the room and its contents. 'Efficiency devices' and 'practical efficiency' were the slogans used to justify and explain the new spatial arrangements of general practice (Arnold and Ware 1953). The constant imperative in the post-war years was 'to economise medical time and skill' (Gillie Report 1963). An editorial in the *Journal of the College of General Practitioners* advised that 'unless the GP organises his working day he will never get through his work' (*Journal of the Royal College of General Practitioners* editorial 1969: 67); work study techniques from industry were advocated as 'a method of using time and effort more economically' (Jeans 1965: 279). Consulting, examining, investigating, diagnosis and treatment were all possible in the same room, advised the Ministry of Health's planning booklet on buildings for general practice, but it was more efficient to have separate spaces (Ministry of Health 1967).

Temporal markers mapped out the new medical space between hospital and home that emerged in general practice but that same space was also subjected to an intensive internal temporal analysis. This is perhaps most salient in the rapid spread of appointment systems: for example, in her 1964 survey Cartwright reported that 15 per cent of patients said their doctor had an appointment system (Cartwright 1967), while in the follow-up study of 1977 this proportion had risen five-fold to 75 per cent (Cartwright and Anderson 1981). Whereas in the inter-war years Hughes' patients waited on their stone bench 'for hours, half a day or longer if necessary', the health centre was meticulously concerned with the placing of patients both in space (by guiding their passage through a highly differentiated series of rooms) and in time through the temporal distributions of the appointment system. In effect, time and space were closely inter-related dimensions:

> The waiting room only needs to be half the size in a practice which works an appointment system than it would need to be in the same practice without an appointment system.
>
> (Whitaker 1965: 267)

In the hospital, time had crept in as an important variable but only in the form of the timetable as a system of ordering staff activity and hospital routines, a system that was self-consciously realized only in the post-war years (Roth 1963). In general, practice there was certainly an element of pacing introduced by the appointment system but the temporal analysis of activity was more pervasive. Time was not a fixed and linear measure but something that was manipulable; time could be used efficiently or inefficiently; time could be conserved or expended. Thus, the temporal space delineated by the appointment system could be further analyzed into 'activities' by means of 'time and motion' studies (Wood 1962), time spent with patients could be titrated against their time needs (Hull 1972) or measured in relation to the social characteristics of patient or problem (Westcott 1977; Raynes and Cairns 1980). Perhaps the Balint group's exploration of what could maximally be extracted from a six minute consultation best illustrates the possibilities for extracting more time from an apparently limited consultation period (Balint and Norell 1973).

Undoubtedly, the rhetoric of 'efficiency' played a large part in this new analysis of time. But to see the changes simply in terms of increasing efficiency would be to miss the fundamental reconstruction of the temporal dimension. Efficiency only became a problem when time became a central concern of post-war general practice. 'Time is the most pressing need for any GP', claimed Townsend in 1962, 'time to spend with and on his patients and time for reflection, thought and reading' (Townsend 1962: 510). It was of vital importance, Batten argued in his James Mackenzie Lecture of 1961, to secure time: 'time to listen, time to think and to talk' (Batten 1961: 18). To that extent, post-war general practice was centrally concerned with an economy of time, with a constant balancing of temporal credits and debits. Time could be wasted in establishing a diagnosis or time could be 'spent' to allow a diagnosis to emerge; time might be conserved in rapid treatment or it might be allowed to pass as a therapeutic tool in its own right (Royal College of General Practitioners 1972). Time, as Harte noted, was working dimension not a linear measure (Harte 1973).

The traditional pathological lesion was revealed at a single point in time: the diagnosis was an event. True, it had a past history and a prognosis, and the diagnosis may, for some reason, have been delayed but these temporal elements were essentially subsidiary to the immediacy of the lesion. In the post-war years, however, when general practice became governed by a temporal economy, time became another dimension, independent of the three-dimensional localization of the lesion that invested illness. Illness was not an event but a process whose context and essential nature was contained within a temporal trajectory. Illness became less the momentary revelation of the clinical examination and more the process of becoming ill, the process of reacting to illness and to treatment, and the process of becoming well. The ultimate event had been the finality of death: in the new general practice, it was successive temporal stages of the process of dying that required negotiation and the expenditure of time (Working Group on Terminal Care 1980). Time became a central attribute of illness; illness was a phenomenon that occupied a temporal space. The emergence therefore of chronic illness as one of the major morbidity problems in post-war general practice marked the crystallization of both illness and time in a common space (Gillie 1963).

As the episodic nature of illness was replaced with a temporal characterization, so the work of general practice was reconceptualized and reorganized. Early studies of morbidity by the Research Committee of the College of General Practitioners noted the difficulty of measuring illness episodes as the end of an episode was impossible to define. The assessment of morbidity in general practice, they concluded, depended on continuing observation (Research Committee 1958). While the research imperative required continuity of observation, patient management demanded continuing care. In 1965 the College, reporting on 'Present state and future needs of general practice', pronounced that the GP 'provided continuing and long-term care' (Reports from General Practice 1965).

An ideology of continuity of care however posed two problems. On the one hand, it was difficult to explain how such a fundamental component of the GP's task had only recently been 'discovered'. This problem was tackled by claiming that it was ever present but that its value had remained implicit: 'Continuity in general practice is so important and so all-pervading that paradoxically it tends to be overlooked' and 'The significance of continuity only became widely appreciated when general practice was first studied as an independent discipline' (*Journal of the Royal College of General Practitioners* editorial 1973: 749). The second problem was that an ideology of continuity of care emerged at precisely the point at which care was being fragmented. Group practice and the health care team meant that the same doctor rarely followed through patients and their illnesses (Aylett 1976). Pinsent, for example, noted in 1969 that 'a stable relationship between patient and doctor is unusual' (Pinsent 1969: 225). The solution to the problem of this fragmentation of illnesses was the medical record.

Post-war British general practice inherited a system of patient records devised in 1920 in the form of a medical record envelope. However, until the closing decades of the twentieth century, very few records were kept in these envelopes or, if they were, their haphazard nature undermined any claims to continuing observation. Indeed, as late as 1978 an editorial in the *Journal of the Royal College of General Practitioners* confessed that 'medical records are the Achilles heel of general practice and reveal the current state of disorganization bordering in some cases on chaos' (*Journal of the College of General Practitioners* editorial 1978: 521).

However, it was not so much that records were bad than that standards by which record keeping were evaluated had been changing. One of the criteria of good practice became a record system adequately filed, every contact recorded, letters stored, summary sheets maintained, regular up-dates carried out, and so on (*Journal of the College of General Practitioners* editorial 1984). 'Only systematic records', claimed Pinsent, 'can replace the series of memories in which a patient's medical history may at present reside, in greater or lesser detail' (Pinsent 1969: 226).

Before records, every patient, every 'contact', was a singular event; there may have been a 'past history' in the consultation and indeed the doctor might have remembered a significant past occurrence but past and present were different domains of experience. The record card, however, revealed the temporal relationship of events so that time became concatenated. Clinical problems were not simply located in a specific and immediate lesion but in a biography in which the past informed the present and pervaded the future. The medical record was another device for the manipulation of time.

As a central element of illness began to be seen as a time dimension, medical intervention aimed itself not so much at a specific lesion as at this temporal space of possibility. Thus, general practice at once embraced and reinforced Surveillance Medicine with its new emphasis on the temporal aspects of illness and biography. As Ashworth noted in a paper on presymptomatic diagnosis published in 1963, 'the role of the GP lies increasingly in the field of preventive medicine is the view of many leading thinkers' (Ashworth 1963). McWhinney's *The Early Signs of Illness* (McWhinney 1964), Hodgkin's *Towards Earlier Diagnosis* (Hodgkin 1966) and a Symposium on early diagnosis in 1967 *(Report of a Symposium on Earth Diagnosis 1967)* all demonstrated this growing emphasis. The role of the GP, observed the government committee set up to examine the organisation of general practice, included the detection of the earliest departure from normal of the individual and families of his population (Standing Medical Advisory Committee 1971). While this approach to prevention and earlier diagnosis might be labelled as part of 'a new approach to disease' (Reports from General Practice 1965), it could also be seen, Crombie suggested, as an implicit part of all general practice:

> The cry for more preventive medicine and presymptomatic diagnosis comes most often those who have clinical responsibility and who do not appreciate that practically everything which the general practitioner does for his patients contains an element of prevention and presymptomatic diagnosis in the widest sense.
>
> (Crombie 1968)

Early diagnosis was a part of a preventive outlook that pushed the identification of illness or its precursors back in time. Equally, intervention in the earliest stages of illness was justified in terms of protecting the future. Thus, early diagnosis and prevention merged almost imperceptibly along the temporal dimension with health education and health promotion. By, as it were, intervening in the past the future could be made secure because the past and future were directly linked. 'We can see prevention', stated the working party on health and prevention in primary care, 'as measuring care with an eye to the future or anticipatory care' (Report of a Working Party 1981: 3).

A temporal economy

The medical model of the nineteenth and early twentieth centuries had pre-eminently defined and given substance to corporal existence.

The disease concept reduced illness to a pathological lesion that, in its turn, was held to be located in the three-dimensional space of the human body. In the new general practice, however, as it shifted from a spatial and corporal to a temporal and biographical model of illness, a novel set of problems, techniques and possibilities was identified. This may have involved the patient-centred consultation and the many trappings of Surveillance Medicine–emphasis on a population constantly 'at risk', chronic illness, prevention, health promotion, anticipatory care, early diagnosis, and so on – but it was the 'progressive' institutional configuration of a transformed general practice that gave them concrete form. The fact that the Old Hospital, that citadel of pathological medicine, began its swift demise at exactly the same time as general practice re-arranged its basic elements was no mere coincidence, nor did it reflect on enlightened health care planning. This was no grand design to drag health care towards the twenty-first century. This was the outcome of a multitude of micro-events – a question to a patient, a sign on a door, an appointment ledger at the reception desk – that together constituted a new way of thinking about illness and a new way of mapping identity.

With the old dichotomy of home and hospital there had always been a domestic space, and a biographical time that escaped clinical surveillance. True, there was the old general practice but that itself was a domestic activity inseparable from the bodies it treated and barely differentiated from lay domestic care. Then, quite suddenly, there was a new general practice, confident, efficient and humane, concerned about biography, subjectivity and community. Just as the domestic body could only be made legible in the neutral space of the hospital, so in the closing decades of the twentieth century it was the domestic biography that was objectified and scrutinized in the intermediate space and temporal vacuum of primary care. The patient's reflexive identity demanded nothing less.

14
Ecce homo

By the end of the twentieth century, the figure of Man presented a very different picture compared to the inert anatomical creature of the nineteenth century, and so did the logic of medicine and pattern of health care. No doubt, other registers can tell a parallel story of Man's creation but few can embrace the brief journey so comprehensively. What other theory or practice can encompass the cells and tissues of the body's depths, the mental processes of the active mind, important institutional configurations, as well as the ecological concerns of late twentieth and early twenty-first century society? It is as if illness provides a magical prism through which to observe both the broad structures and the fine detail of the many changes in the identity of Man over nearly two centuries.

In the nineteenth century, the new medicine based on the search for the pathological lesion not only signalled a new way of thinking about and dealing with illness, but also marked out the three-dimensional outline of the now familiar analysable human body. From the novel conceptualization of illness that linked surface and depth, through the techniques of clinical method that celebrated the volume of that body, to its unencumbered observation in the hospital ward, medicine defined and redefined a discrete corporal space. In other words, the birth of Man in the nineteenth century occurred alongside a fundamental reconstruction in clinical practice in which pathological medicine centred on the hospital demanded the analysis of bodies. Individual physical identity was thereby forged in the practical anatomy of clinical work.

The corporal manifestations of illness that marked the new pathological medicine came to dominate the nineteenth century and succeeded in maintaining their ascendancy during much of the twentieth. To be sure, the techniques for identifying the hidden lesions of the body grew more sophisticated. The addition of laboratory investigations to the repertoire of clinical indicators of disease towards the end of the nineteenth and throughout the twentieth centuries increased the power and intensity of the clinical eye's ability to peer into the depths of the body's density. Yet while clinical investigations in the form of X-rays, pathology reports, blood analyses, etc, marked an extension of the technical apparatus of medical procedures, it did not challenge the underlying spatial arrangement of illness or the logic of clinical practice. Experience and illness were still linked through surface and depth, inference of the true nature of the lesion still dominated medical thinking, and the hospital still – indeed even more so – remained the centre of health care activity.

As the figure of Anatomical Man materialized so the complexity of his inner form and outward separation from nature demanded explanation. Surely, the hand of God could not have created this corporal space in a single day, even less ensured its absolute separation from the work of the previous days? The Genesis story could claim that following his transgressions Man was expelled from the natural order but the great quest was always for return. The new narrative had to recognize that Man was prized from nature and only existed in separation. True, Man was reunited with nature at the end of life though this was hardly redemption requiring elaborate hygienic rituals to ensure that the living were not contaminated as the corpse slipped away; it was not a heavenly chorus that welcomed the dead but proper and complete physical decomposition in the bosom of the earth. Man needed a new creation story and Darwin provided it.

The new Darwinian story described the origins of the anatomical body not in terms of the contemporary techniques that were even then modelling a Man distinct from nature, but in a long-distant past. At that moment, Man had both an identity and a history, and the search for those mythical origins in the rocks of time could begin. Yet while the Darwinists hunted for their fossilized evidence that would explain the present, Man was already changing faster than their theories could ever allow. How ironic that the Lamarckian heresy, the real challenge to Darwin's triumph, was so decisively defeated.

The early twentieth century witnessed the first faltering steps of this new creation. Movement meant interaction with neighbouring bodies and the subsequent discovery of the Other as identity became defined by its relationships rather than its anatomical integrity. The moving interacting body of the early twentieth century was waking and stressing its autonomy not only as an object but also as a subject. Disease was released from its prison of the body and spread to the new social spaces of identity.

The individual clinician had been one of the first explorers in the new continent of psychological and social spaces as revealed by patients' words. Then other strategies for penetrating into this new world rapidly emerged, the most important of which was the survey. Surveys that simply counted had a long history though they grew in number and extent during the nineteenth century. In these surveys, respondents were treated simply as reporters of the world external to them. The idea that survey techniques could also be used to explore the vast unknown geography of the inner world only began to emerge in the mid-twentieth century alongside the clinical experiments that elicited the early signs of the patient's mind.

The addition of subjectivity and reflexivity to corporal identity was therefore an accomplishment of a certain form of clinical practice that engendered the autonomous whole-person. Symptoms, for example, were accorded a new status as reflecting not the character of the lesion so much as the identity of the patient. This meant that there were two parallel transformations in medicine: one was in the detail, procedures and organization of everyday clinical practice, the other was in the theoretical model of illness that underpinned and legitimated those practical displays. The medical model that emerged in the late twentieth century was therefore a very different one from that of the mid nineteenth century, which, in its turn, had been revolutionary in its relationship with that of the eighteenth century.

Corporal identity had been dependent on a form of medicine that kept illness and health separated by a conceptual and practical gesture; the dissolution of the boundary between health and illness under a regime of Surveillance Medicine implied a loss of that anatomical detachment. Identity then began to crystallize in a new temporal and multi-dimensional space whose main axes were the population – within which risk was located and from which risk was calculated – and a temporal space of possibility. The new identity was to be found in the shift from a three-dimensional body as the locus of illness to the four-dimensional space of the time-community. Its boundaries were the permeable lines that separated a precarious normality from a threat of illness. Its experiences were inscribed in the progressive realignments implied by emphases on symptoms in the eighteenth century, signs in the nineteenth and early twentieth, and risk factors in the late twentieth century. Its calculability was given in the never-ending computation of multiple and interrelated risks. A model of illness; a clinico-social practice; but also a dream of healthy living that was celebrated everywhere, from the promise on the margarine packaging to the slogans in the lifestyle magazines, from the confessional techniques of an enlightened medicine to the health promotion activities of a transformed public health.

In the late twentieth century, Darwinism needed revising to take account of the fact that the fossil evidence of Anatomical Man only described a long dead figure. Behaviour, mental processes, subjectivity, agency, and so on, needed accounting for and incorporating into the old story of origins. A flurry of interest in animal behaviour in the 1950s and 1960s heralded a concerted attempt to understand the origins of human behaviour and at the same time return Man to his natural ecological niche (Tinbergen 1953; Lorenz 1966; Morris 1967, 1969). Animal behaviour could be projected into the space of Man's identity and his place in the natural order finally settled. But how could a theory that fixed Man's identity in a distant past explain fully the dynamic and changing object of the twentieth century? Certainly, it had worked well for the inert figure of Anatomical Man but the theory needed constant revision to explain a rapidly changing multi-dimensional identity. In the end, it was Darwinian theory that needed to evolve as it adapted to maintain a credible creation story.

This leaves open the possibility of other creation stories that might account for the great journey from proto-Man through primaeval body to subjective, acting, reflexive self, from the fixed solidity of anatomical identity to the precarious and shifting spaces of late twentieth and early twenty-first century existence. One such analysis involves a political geometry that construes identity as the resultant of lines, planes, axes and forces that together establish the necessary spaces for the possibility of Man.

Political geometry

The way in which identity was fashioned through lines and spaces can be outlined through the four regimes of public health that succeeded each other during the nineteenth and twentieth centuries. First, there was quarantine that drew a line between places. Secondly, sanitary science guarded a line between the body and its natural environment. This was followed by the regime of interpersonal hygiene that persuaded those same bodies to maintain a line between each other. Finally, there was the new public health that deployed its lines of hygienic surveillance everywhere throughout the body politic.

Each line, in its turn, delineated and thereby created a new space – at once both the space of illness and the space in which identity could materialize. Quarantine identified masses bound by a geographical place that had to be kept separate; it had been a totally anonymous technology that almost inadvertently caught people – sometimes few, sometimes many – within its circumscribed net. Then sanitary science dissected the mass and recognized a separable and calculable individuality in the form of anatomical/corporal space in the former crowd – though not yet the singularity of individual difference. These individuals were allowed their free passage so long as they subscribed to continuous surveillance and safeguards to separate everyone from the hazards of external space.

With quarantine regulations the police, troops and magistrates had been urgently mobilized to protect the common safety of the state. Sanitation was concerned with longer term and very different goals: 'The aim of Sanitary Science is to prevent disease, preserve health and prolong life' (Guy 1870: 5). Yet this was simply the means to a more important end, namely 'to maintain the whole people in the highest efficiency for the labours of peace or the struggles of war' (Guy 1870: 5) No longer was public health synonymous with passive protection but became a productive strategy along with every factory and industrial technique of the nineteenth century.

The body that emerged like Frankenstein's monster from the clinical and sanitary science laboratory of the nineteenth century was then further transformed by the techniques of inter-personal hygiene and social medicine. These techniques 'discovered' and maintained a psycho-social space in which crystallized a new facet of human identity in the form of psychological difference. This can be seen in the extension of psycho-dynamic ideas into clinical thought early in the twentieth century and in the contemporary study of aspects of individual differences, such as intelligence, personality and attitudes. In similar fashion, in the second half of the century, the techniques and focus of the new public health began to fabricate yet another space. The contours of this newest space outlined a novel configuration of the reflective subject in the setting of social, economic and political activity.

In each case, the lines of separation divided an existing space into two or more new spaces. In the case of quarantine, the line could be drawn on a map, but later lines that ran through a proliferating multi-dimensional space make any pictorial representation more difficult. The three-dimensional body of sanitary science, in part, could be sculpted by means of three-dimensional diagrams or other aids to seeing volume, such as transparent overlays or cross-sectional pictures, but the multiplying spaces of the twentieth century were more difficult to map on the two-dimensional page. This does not make them any less 'real' than the geographical landscapes of systems of quarantine or the tissues and organs of the body enclosed by sanitary science, though they tend to present themselves as dimensions that are more abstract. Even so, they can be mapped by the same techniques of analysis that were implicated in their creation, a process of following the different axes and attributes laid out and made real by the practical work of everyday medicine.

These various spaces hold the key to identity. The 'primary' division of a space defined both what was and what was not Man; but then the space of Man itself sub-divided across many planes thereby creating a multi-faceted identity. The 'problem' of identity is not therefore an old question – though it can become so with a little retrospective projection – but belongs to the very recent past when spaces, axes and planes multiplied. What is part of Man and what is not thereby loses the clarity it seemed to possess little more than a century earlier. In the end, these indistinct boundaries together with the search for coherence cross these many planes of existence culminated in the gradual dissolution of that once proud nineteenth century figure.

While identity can be portrayed in terms of shifting and dividing spaces, the process of 'space formation' itself can be illustrated with a closer examination of the lines that delimited the spaces of existence – corporal, psychological, social, etc. Thus, the line of separation of quarantine, which had no depth, volume or permeability, might be seen as the most simple. It was essentially a mechanism of exclusion. Things stayed either side of the line without recognition or analysis of separate bodies or individualities.

In sanitary science, the line was more complex. Indeed, it was hardly a line as there were various volumes of the body – such as the mouth – whose exact status and relationship to the line were unclear. Moreover, the line was permeable in that objects such as air, food, water and bodily wastes continually crossed it. The apparent line of sanitary science was more a space, thin and contoured, but nevertheless distinct. This space – where it materialized from the simple line of body boundary – was a physical volume with three-dimensional properties; but while those boundaries of skin, mouth, nostril, colon, and corpse were anatomically defined, their exact topography was not so clear. It was therefore only through careful and constant vigilance that their hygienic properties could be ensured and their bounded existence maintained.

The apparent lines that separated different bodies within the strategies of interpersonal hygiene and social medicine established volumes that extended beyond the three dimensions of the simple physical body: it was above all else a psycho-social space within which actualized the objects discovered by the contemporary new sciences of psychology and sociology. There were instincts and attitudes, behaviours and actions, statuses and roles. It was through the creation of these spaces that the various characteristics of individuality became clustered around the physical body that had earlier been defined by sanitary science.

The new public health extended the psycho-social space of interpersonal hygiene through the discovery of danger everywhere. Bodies were under threat from pollutants and contaminants in the food chain, in the air, and in the water; minds were under threat from Martians, communists, fanatics, perverts, television, teachers, and cinema. These dangers came not from 'nature' but from Man's corruption of that wholesome space. A vast network of observation and caution was therefore deployed throughout late twentieth century society, involving not only the vigilance of the public health authorities but the attempted involvement of everyone in the surveillance task, particularly through global strategies such as 'Health for All' and health promotion. This perpetual guard against threats from the interactions of others generated a sort of 'political' consciousness that might be described as reflexivity, a reflection about self. It was only through constantly reflecting on the dangers that menaced self that a new identity was sustained. It was the thinking, acting subject that was both object and effect of the new public health in all its various manifestations.

The creation story of shifting lines and spaces finds its own origins in the period that it tries to explain. Durkheim made the link between lines, spaces and the social in 1912. In *The Elementary Forms of the Religious Life*, he argued that the fundamental basis of social classification was religious in origin: 'By definition, sacred beings are separated beings' (Durkheim 1912: 299). Famously, Durkheim argued that the first and most fundamental sacred region was the social itself. The line drawn between sacred and profane worlds was based on the principle of exclusion, such that 'an abyss' or 'a sort of vacuum' separated these worlds (Durkheim 1912: 318). Thus, while Durkheim clearly saw a barrier between the two worlds he also hinted that it had depth and volume, though of an incalculable sort. Moreover, just as Durkheim described the emergence of a new space of possibility from between the established categories of a classification system, the great nineteenth century framework of sanitary science was just giving way to an interpersonal hygiene that identified and monitored a new 'social' space. Was it coincidental that Durkheim's writings celebrated the primacy of the social?

On a large-scale map of Europe, the battle positions of the opposing armies in the Great War of 1914–18 would be drawn as a narrow space defined by the trenches of both sides. This space was called No Man's Land. The term No Man's Land had been used five centuries earlier to describe an area of ground to the north of London that lay outside the patterns of customary ownership. This land was used for public executions. Similarly, the No Man's Land of the Great War was not only contested ground but was also a narrow geo-political space, unique throughout a Europe of sovereign states in being outside the established order. Behind the forward lines, sovereign governments exercised power to maintain a juridical order based on a civil penal code. No Man's Land, however, did not offer a power vacuum: it was governed by a massive and capricious coercive force in which sudden death and destruction were the constant risks.

Durkheim's 'abyss', No Man's Land and the public health regime of inter-personal hygiene have more than contemporaneity in common. The nineteenth and twentieth centuries have witnessed a redrawing of the map of public health surveillance as hygienic attention, starting from the line of the cordon sanitaire, moved to body boundaries, then seemed in close up to dissolve its focus and reveal a potential space of inchoate dimensions and properties. It is within this space – a sort of political No Man's Land – that individual identity materialized and was refashioned. It is as if those trenches that defined the dangerous space at the beginning of the twentieth century did not disappeared with the coming of peace, but rather rolled back across the landscape, revealing a vast space of limitless dimensions in which the social and subjectivity would crystallize.

A half century after Durkheim's work on classification systems, Douglas, in *Purity and Danger*, identified a number of objects and social practices contained within the space that Durkheim had only described as an 'abyss' (Douglas 1966). Indeed, in Douglas's view this intermediate space functioned to maintain the classification system by acting as a residuum, as a home of last resort for any anomalies. But, crucially, this dumping ground posed its own dangers and threats. Again, Douglas was writing on the threshold of a major change in the system of public hygiene. At the time that Douglas called attention to the incipient dangers lurking within the spaces lying outside of and between the main categories of the extant classification system, public health was beginning to embark on the great crusade to promote an ecologically conscious hygiene that recognized the existence of danger everywhere.

The parallels between the sociological/anthropological writings of Durkheim and Douglas and the prosaic routines of everyday public health are striking. The space of danger defined the space of existence yet lay outside of it. The anatomical body stands as an exemplar of that technique as do the proliferation of deviancy models about the time that Douglas was writing that stressed the role of the out-group in defining the core nature of the in-group. Towards the end of the twentieth century, however, this creative relationship was reversed. Whereas for Durkheim and Douglas the line/space that separated categories delineated contemporary identity, the space opened up by the new public health and the reflexive surveillance that underpinned it was both the space of danger and the space of identity. At the beginning of the twenty-first century, the order of social life is revealed as having its origins in disorder and chaos.

What then were the characteristics of the lines that divided up the political map? Were they lines or were they spaces, and what mechanisms of power maintained their morphology? Was it the ordered space behind the lines that protected health, or the abyss between the lines that produced both the object and subject of wholeness? The image of the Rubicon, of a simple political geometry, has dominated social classification for too long: surely, it is time to examine the currents in that dangerous little stream.

Through the prism

This creation story is a nominalist account of identity. There is no essence of Man; he is only a coalescence of different attributes at any one point in time. Man can be represented by an externally defined space, the residual volume of spaces that surrounded him. Yet such spaces are not restricted to the three-dimensionality of the physical world but embrace the plurality of planes that appeared during the twentieth century. This means that Man can never be known in his entirety. In part, the target is a constantly shifting one; in part, Man was always seen through a prism. Each of the shifting spaces of identity were perceived, read and analysed through a lens that was itself always changing. Accordingly, there was no reality to identity, only the remarkable flickering object seen through the perpetual prism of both illness and death that projected a story of life.

This preternatural prism reversed the polarity of life and death, and of perception and agency. For nearly two centuries, a single myth prevailed that death intervened to end life. But the prism showed that both illness and death preceded life, defining it, constituting it, authenticating it, sustaining it. It was through illness that

Man came into existence and through the changing nature of illness that his identity was transformed. The prism allowed the observing eye to see an independent world beyond, yet one that was governed by the reflecting and refracting surfaces that radiated and fired the images of Man's existence. It was not the acting Man who perceived but perception that invented the acting Man. It was not Man who derived the models of illness through which he could be analysed but the models that created the thinking Man. It was not Man who established the institutional setting – hospitals, community care, etc. – that allowed his wholeness to be maintained but those same institutional settings that generated a very wholeness. So, the question of the mystery of Man can be recast: not 'who is Man?' but 'who – or what – constitutes him?'. Who is this anonymous observer? What is the source of the disembodied eye in its unrelenting gaze through the prism of perception?

15
Identity of the Observer

During the 1960s, the shift of clinical activity from the home of the general practitioner (GP) to the health centre marked the strengthening of the boundary between the GP's own domestic/private life and general practice work. This barrier between private and professional lives was further consolidated by changes in the temporal organization of practice activity. For the traditional inter-war and earlier single-handed GP, time was not fragmented: there were no regular hours or specific times designated as 'off duty', and her patients who sat and gladly waited, perhaps for half a day, to see her seemed equally oblivious to temporal pressures or boundaries. By the 1950s, however, there was a new recognition of the importance of time as a factor in clinical activity in that morning and evening surgeries, together with specialist services such as antenatal clinics were often allocated specific hours, though the remaining practice occupied the undifferentiated time, especially for home visiting and being 'on-call'. Then, a few years later, in the setting of the health centre, the GP's time was completely transformed as it was subdivided and actively managed.

Besides placing clinical work within a temporal frame, the new pattern of fragmented activity also had an impact on the role of the GP. 'Off duty' periods, as shown in the spread of rotas and deputizing services, identified and separated times when the GP was acting as a GP and times when she was not. Equally, the contemporary development of a formal 3-year training period to replace a system of casual entrance also served to raise a temporal barrier between identity as a hospital-based practitioner and life as a GP (Horder and

Swift 1979). In effect, a 'GP identity' began to be created by drawing a line within the GP's former professional role so as to mark a separation between a GP's life and non-GP (hospital) doctoring and then drawing another line between the GP as professional and GP as private person.

Although seemingly minor, these organizational changes in general practice that started in the late 1950s and early 1960s signified a remarkable development in the emergence of identity. For over a century medical analysis had focused on the patient – at first a body, later a mind, both subjective and reflexive. Medicine had scrutinized the patient, eliciting new facts, cajoling the appearance of new phenomena, and crystallizing facets of identity; but always the medical gaze had been disembodied. A great medical eye of surveillance without a tangible presence and without an identity beyond its power to observe, describe and monitor. But here, at the very moment of triumph for patient subjectivity, a new identity began to be distilled, that of the once hidden observer, the eye behind the prism of medicine. It was as if the clinical gaze that for over a century had opened up the space of patienthood began to turn inward to examine the proto-space of identity behind the eye of surveillance. Who was this shadowy figure?

Identity of the doctor

Curiously, the emergence of the doctor's identity did not follow the same pattern as in the patient, namely a body followed by a psychological identity. Instead, the mind of the doctor preceded the appearance of her body. However, the process through which both the eye of medicine was given 'personality' and then became embodied can best be described when it addressed its primary task, namely engagement with the patient.

There had been some writing on the interchange between doctor and patient in the inter-war years. The new psycho-social space of the patient that emerged during the early decades of the twentieth century necessitated some reassessment of the relationship. Brackenbury, for example, observed in his book, *Patient and doctor* of 1935, that 'the relationship between patient and doctor is not merely between two persons, but between two personalities' (Brackenbury 1935: 74). Even so, the personality of the doctor was not very visible. When Brackenbury went on to observe that it 'is never the body which is out of health, but always the complete being' it was the patient rather than the doctor that was his object of attention (Brackenbury 1935: 74). Indeed, he argued that the patient could anticipate what might be called a professional role from the doctor – knowledge, skill, carefulness, judgement, sympathy, understanding, moral character, and ethical conduct – thereby subsuming the doctor's identity to that of her professional group.

Recognition that there was a subjective interplay between doctor and patient emerged in the immediate post-war years. Parsons' 1951 writings on the doctor–patient relationship dealt with the mutual and reciprocal obligations of both parties but continued to construe the actors in their respective social roles (Parsons 1951). A few years later recognition that the doctor's own identity might be compromised by constant elicitation of the patient's began to be recognized. In 'The doctor, his patient and the illness' (Balint 1955), Balint described the traditional medical 'Apostolic Function' whereby doctors' behaviour tended to define the boundary between those illnesses that could be admitted as legitimate objects for their attention and those that could not. He claimed that by the twentieth century the Apostolic Function had achieved remarkable success in persuading patients to disrobe and allow a physical examination: 'what otherwise would constitute a serious violation of modesty' (Balint 1955: 685). According to Balint, in the 1950s the Apostolic Function was beginning to liberate a territory called 'the mind' for diagnosis and rational therapy just as it had earlier successfully made available 'the body'. Indeed, Balint-trained GPs found that patients 'were not only willing to undergo, but demanded, a psychological examination' (Balint 1955: 685) – yet further evidence of the mid century reconstruction of the medical encounter and reconceptualization of the patient as powerful new concepts permitted doctors to make interpretations that defined a psychological reality lying beneath patients' presentation of their problems.

Whereas patients could to some degree control the penetration of the clinical gaze in a physical examination by detaching 'themselves' from 'their bodies', in a psychological examination this became vastly more difficult, as all their behaviour, overt and covert, was potentially legitimate material suitable for decoding by the doctor. In the new 'personal' doctor–patient relationship, the boundaries were complex and indistinct and the possibilities for transgression correspondingly multiple and ill-defined. Balint therefore warned against an overzealous use of psychological interpretations that might rob patients of their symptoms in a kind of 'psychological tour-de-force, which is really a violation of a person's private life' (Balint 1955: 685). Indeed, much of his argument was concerned with the ethical dilemmas that arose in each encounter with an individual patient: deciding what should be seen and what left unseen; what should be said and what left unsaid; what should be organized and what left unorganized. GPs had to make difficult choices for which their professional training had ill-prepared them.

The management of the uncertainty of the consultation meant that the individual doctor's view, hitherto indistinguishable in medical discourse from the collective impersonal view of the medical profession as a whole, became subtly complicated by a degree of subjectivity. The rhetoric was that of altruism and objectivity, but the effect was to open up an area of ethical discretion that extended beyond the essentially technical concept of clinical judgement. How was the profession to introduce the sensitivity of subjectivity into the clinical gaze without sacrificing the power associated with its reputed objectivity? How was it to engage beneficially the doctor's self, without dangerously engaging her self-interest? And how was it to render the doctor-patient relation as a space open for inspection, while preserving its intimacy? One solution can be found in the post-consultation analysis pioneered in 'Balint groups'.

If the post-mortem examination of the body by dissection was the defining activity of the anatomo-clinical method, then the corresponding activity in the foundation of the 'doctor-patient relationship' was a form of consultation inquest. Balint groups involved a number of GPs meeting to discuss a recent consultation presented by one of their members. The material examined was not a dead body, but a narrative about a dead social interaction presented by a witness (who was also a protagonist). An important role for the facilitator of the group was to ensure that discussion remained focused on the space between doctor and patient rather than straying too far into the personal psychic life of either GP or patient. These conventions, which were adopted ostensibly in order to meet the requirements of scientific objectivity, also had another less overt effect of splitting doctors' subjectivity into a public aspect, admitted by the group, and a private aspect that was excluded. Thus as Balint maintained, the Apostolic Function had 'private and public sources and aspects ... most of them are, so to speak, private; they are expressions of the doctor's individuality; and though their importance is obvious, I shall say nothing more about them' (Balint 1955: 685).

In effect, the genie of medical subjectivity was not to be released, but rather installed into a larger and better-furnished bottle, designed to keep the professional and the personal hygienically separate. The boundaries that had been established to protect the patient (from abusive interpretation) and the doctor (from cross-contamination between personal and professional emotions) effectively delineated a new social space, a space between individuals in which their relationship could safely be examined. Objective scientific scrutiny of this space would ideally require there to be a detached observer in the consulting room, but 'the presence of a third person, however tactful and objective, would inevitably destroy the ease and intimacy of the atmosphere. Such a third person would see only an imitation, perhaps a very good imitation, but never the real thing' (Balint 1964: 3). Nevertheless this did not prevent a succession of studies, particularly by social scientists, that attempted to open up the space of doctor-patient interaction to a wider scrutiny (Stimson and Webb 1975) and tried to objectify through a scientific rigour the subjective encounters that the interaction subsumed.

In Balint's system, the objective gaze was provided by the group, removed from the consultation in space and time, but linked to it by the unelaborated testimony of the presenting doctor. This doctor had a strangely split role, being on the one hand the bearer of a narrative of subjective experience (analogous to a symptom) and on the other a member of a group whose meticulous examination of that narrative provided an objective version of events (a diagnosis). This doctor's gaze thereby became fragmented, multiplied and displaced. A potential strain was set up between her point of view as a subjective individual and that held as a member of a professional group, between her identity as an actor in a social field and as an objective observer of her own performance. One might conclude that Balint's 'tactful and objective' third person had indeed been smuggled into the consulting room, concealed in the newly fractured identity of the professionally self-conscious doctor.

The tension between professional identity and subjective self-awareness was reflected in contemporary analyses of doctors in their collective role. The anonymous eye of medicine might have been exercised by individual practitioners but was a characteristic of the medical profession as a collectivity. During the late 1950s and 1960s a new discourse on professions and professionalization arose to explain the historical emergence and social place of this unusual occupational group. Finding historical precedents in Durkheim's idea of professional guilds as protection against anomie (Durkheim 1893) and in Wilson and Carr-Saunders inter-war apologia for professional status (Carr-Saunders and Wilson 1933), the new literature stressed the importance of precisely those ethical principles that prevented the collapse of professional into personal identity (Goode 1960). And it was those ethical principles–altruism, community orientation, a service ideal – that justified the profession's high social status and its immunity from outside scrutiny. Besides, who could look directly into the eye of the medical Medusa? Who could return the transformative stare of medical observation?

In 1970 Freidson published *Profession of Medicine* in which he rejected the 'ethical' analysis of medical practitioners as simply reflecting professional self-interest (Freidson 1970). Instead, he argued that it was the profession's right to define and redefine illness that was the basis of its professional power, and that right had not been freely given but had been expropriated. This reassessment of the medical profession as a self-interested group with less than fully altruistic motives opened up professional status to critical analysis at exactly the same time as the professional role was beginning to become entangled with personal identity. Clinical practice, which had provided the locus for the observing gaze since it started its great creative mission in the nineteenth century, was itself beginning to be dissected: the prism of observation was slowly turning its refractive surfaces towards the collective eye of medicine.

Medical reflection

The separation of private and public in the doctor's identity was a temporary affair. The very identification of these separate realms of conduct and experience marked the beginnings of an individualized identity for the shrouded medical eye that had for over a century had observed dispassionately the patients's body and latterly the patient's mind. A reflexive gaze began to turn on the doctor herself. An important part of this process was the creation of a vocabulary and grammar of doctors' subjectivity, embedded in empirical methods devised for studying and teaching the doctor-patient relationship.

In their *Treatment or Diagnosis* of 1970 Balint and his colleagues introduced a distinction between traditional 'illness-centred' medicine and a new 'person-centred' approach, thereby formally recognizing a potential shift of focus in the GP's field of attention from an 'illness' to a 'person' (Balint *et al*. 1970: 25–6). A few years later in their study *Doctors talking to patients* (Byrne and Long 1976), Byrne and Long presented for the first time verbatim transcriptions of doctors' speech in their surgeries. In this raw material the authors discerned patterns of verbal behaviour which they grouped into 'styles' of consulting, defined as either 'doctor-' or 'patient-centred'. It was as if the voices of the doctors in the study had been transformed into text, distilled and reconstituted into two iconic synthetic voices, each representing an idealized point of view. Byrne and Long's 'doctor-centredness' was the voice of Balint's 'illness-centred medicine', while 'person-centred' medicine was articulated by the new voice of 'patient-centredness'. Terms that had previously defined the kinds of object examined by doctors (i.e. an illness or a patient) had mutated into descriptions of the point of view adopted by the examiner herself and, by implication, something about her character – her way of being as revealed in her way of seeing.

A key finding of Byrne and Long's study was that each doctor appeared to have a strong tendency to stay within a narrow range of styles: 'when a doctor develops his style there is a possibility that it can be a sort of prison within which the doctor will be forced to work' (Byrne and Long 1976: 118). This finding actually consisted of two parts: the prevalence of stylistic rigidity *per se*, and the fact that the predominant style was one of 'doctor-centredness'. The authors clearly did not consider the doctor's inflexibility to be a problem in itself since a 'doctor-centred' style was regarded as adequate for the management of acute organic pathology; and yet it seemed that there was a need for a range of behaviours that would meet the new psychosocial problems presenting in practice. Rather than probing further into the reasons for doctors' behavioural inflexibility, Byrne and Long proceeded directly to considering how best to supply the missing behaviours. Acquiring this expanded repertoire, however, did not imply a liberation from former constraints; rather it represented a means of preserving the doctor's ability to respond in an habitual and stereotyped way in the face of a greater variety of situations: adding more rooms to the prison of style rather than escaping from it.

'Patient-centredness' might well be viewed as simply an updated form of 'doctor-centredness', informed by a need to elicit the patient's subjective world. Yet disclosure of biographical information was, according to the patient-centred doctor's view, presented as self-evidently in the patient's interest. In effect, the new rhetoric of the doctor-patient relationship, as well as enabling GPs to differentiate themselves from their 'illness-centred' medical colleagues, allowed them to preserve a view of themselves as particularly humane doctors. This was achieved by abstracting the rhetorical values that were seen as being attached to traditional family doctoring and importing them into the reborn specialty – even though, ironically, general practice was contemporaneously moving out of its traditional location and into the impersonal health centre. Thus 'entering the patient's world' was to be carried out on a symbolic level in the doctor's consulting room rather than by visiting the patient's home, while 'seeing through the patient's eyes' really meant adopting a new official language that sought professionally to re-define what it was that patients were supposed to be expressing.

The new approach to practice had provided the 'patient-centred' doctor with powerful tools for exploring and describing her patients' world, without the corresponding means to call into question the nature of her own. The result was the creation of a strange paradox: a perceptive, 'personal' doctor who was not yet, herself, a person; a subjectivity without a subject. The vacuum soon began to be filled by the formation of a doctor's mind as an analyzable distinct phenomenon. Training in the new general practice therefore came to involve the development of the doctor's own psychological self:

> I am becoming more aware of personality and feelings, and of the ways in which I interact with patients. I am beginning to recognise the ways I can use different facets of my personality subtly to influence the behaviour patterns of others.
>
> (Stott 1983: 59)

As with so much of the doctor's psychological identity, her suffering emerged from an imperative to engage with the patient. Earlier in the century, pain had been viewed as the archetypal symptom, the direct record of a lesion based on stimulation of sensory nerve endings. In parallel with the growing subjectivity accorded the patient in the late 1960s an increasing importance was ascribed psychological processing in the perception of pain (Melzack and Wall 1965). In his later manual on clinical method published in 1976, McLeod went so far as to claim that pain was a 'purely subjective complaint'. This meant that it needed skilful handling in the consultation – with important implications for the doctor's own identity:

> Its subjective nature is such that it is only through personal experience of pain that a doctor can have insight into the meaning of the descriptions given by patients.
>
> (McLeod 1976: 9)

Reliance on self-experience to grasp the meaning of symptoms was also found in a book published in 1977 of sociologists' accounts of their own illnesses (Davis and Horobin 1977). The authors criticized previous sociological studies of illness as being 'too formal, objectified, detached and scientifically rigorous ... each illness experience and encounter with organized care is unique'; it seemed that the only means of transcending the interpretation of meanings, of achieving authenticity, was to observe not the illness of others but the illness of self. The patient's view, caught in a dense web of subjectivity, was becoming the template from which the doctor's view could be constructed.

The formalization of a world of experience and identity for the doctor transformed the nature of the encounter between doctor and patient. In many ways, the notion of a doctor-patient relationship failed to capture the essentially contingent and precarious scenario contained within the term doctor-patient interaction. As McLeod observed:

> In addition to the patient's response to his problem the interactions between the patient and the doctor have also to be considered. This relationship is very complex as a result of the interplay between different personalities in potentially unstable situations.
>
> (McLeod 1976: 9)

Browne and Freeling, drawing on transactional analysis, likened the interaction to a game in which doctor and patient could each adopt a series of different roles (Browne and Freeling 1976). In the same year Wadsworth and his colleagues, also recognizing that the encounter between doctor and patient was a two-sided affair, set out to investigate the rules, routines and procedures that doctors and patients use to organize consultations (Wadsworth and Robinson 1976). Doctor and patient were now two personalities circling around each other engaged in a process of mutual constitution.

Embodiment

The only thing missing now was a body for the disembodied gaze. A hint of things to come can be found in Byrne and Long's study in an early oblique recognition that the clinical gaze not only illuminated but also originated in a body. In a humorous aside to their discussion of the meaning of patients' verbal behaviours, the authors commented: '... most doctors agree that when they hear these words they start to suffer unpleasant feelings in various parts of their anatomy which have no clear organic cause' (Byrne and Long 1976: 14).

In his photo-essay of 1967 about the work of a country GP, Berger caught the faint presence of the doctor's body in grainy black and white images of landscape and suffering (Berger 1967). Sometimes the photograph showed a hand, or an arm, or a figure caught momentarily from behind; sometimes a nameless face appeared without an identifying caption: how could the face of the doctor be distinguished from the face of the patient? The effect was of an anonymous professional carrying out a professional's (rewarding) work. This 'fortunate man' was only a silent witness to the land, sky and suffering; but her own suffering was barely a decade away.

During the 1980s the old 'doctor-patient relationship', with its resonance of psychoanalysis and long-term pastoral care, began to be replaced in general practice discourse. The new object of interest was 'the consultation', an interaction of relatively short duration that lent itself to recording, transcription and analysis. In 1984 Pendleton and his colleagues described an approach to teaching consultation skills in which behavioural science and technology were to be used non-didactically to train doctors in effective communication. They claimed 'we are not prescribing consulting methods – we do not want all doctors to consult like automata ... we want doctors to define what they wish to achieve in their consultations and to be able to bring this about in their own way' (Pendleton *et al.* 1984: 40).

The way that doctors were to learn new consulting skills involved them being provided with feedback on their performance, usually in the form of a recording, which by this time was likely to be on videotape. The learning would occur by defining the goal of the consultation, then using the recording or other feedback to compare the behaviour of the novice or ineffective consulter with that of 'individuals who are able to achieve the goal'. This comparison would thereby allow the novice to identify and learn more effective behaviours. However, as Pendleton and his colleagues pointed out, the middle phase in the teaching of social skills (comparison with the actual performance of an effective practitioner) tended to be omitted so that teaching 'becomes based on the assumptions of the expert' as to the elements of her own behaviour that have been effective (Pendleton *et al.* 1984: 26). It would seem that an apparently permissive attitude, which claimed not to prescribe the actual behaviour of doctors in the consultation, concealed less overt processes of person management.

Where the traditional medical model had dictated stereotyped forms of interrogation and procedures of examination that were to be used by doctors to achieve a diagnosis, the new, more 'personal' model required the use of a new instrument in the form of the doctor's self. The self that the doctor brought to the clinical encounter, however, was not a naive self, but self honed and calibrated to detect certain categories of phenomena. Far from liberation of the doctor's inner nature from the shackles of ritual and convention, this was actually a process of bringing the doctor's formerly private self into the clinical field where it too could be subjected to the effects of clinical discipline. The message carried by the new methods of training in consultation skills was not so much 'know thyself' as 'invent thyself'.

The objective correlate of this process was the creation of a representation of prescribed style–an image of the doctor on a television screen which became not so much a means of learning, as an end in itself: the creation of a doctor, by a doctor, for view by other doctors. The doctor, having acquired a voice in Byrne and Long's study, now developed a body, albeit a virtual image displayed on a screen; a communicating object strangely dissociated from a watching 'self' who was to observe, learn from and control it.

In this new culture, doctors were required to relate not only to patients, but also, through the use of video, to an image of themselves in the process of relating to patients. Furthermore, since video recordings were reviewed with peers and trainers, doctors now also had to relate not only to their colleagues' appraisal of the behaviour recorded on the tape but also to their colleagues' appraisal of their own self-appraisal. This represented a kaleidoscopic expansion of points of view and frames of reference. Even as the image of the patient as a whole person became clearer and more comprehensive, so the identity of the doctor became increasingly hard to locate.

In 1987, Neighbour's *The Inner Consultation* found the 'doctor-as-subject' in a rather fragmented and dissociated state (Neighbour 1987). Neighbour described an 'internal dialogue' that frequently disrupted the flow of consulting with confusing self-observation and criticism. The increasing ability of the doctor to analyse and organize seemed paradoxically to paralyze her ability to respond. The solutions offered in Neighbour's text involved re-framing of the doctor as a kind of embodied information processor, the structure of which represented a new spatialization of the doctor's subjectivity and the arrival of perhaps the first fragments of an anatomy of the doctor's subjective body.

The intuitive, responsive aspects of the doctor were held to be located in the right hemisphere of the doctor's brain, and the rational, organizing aspects in the left. The five 'checkpoints' that were to be reached in the course of the consultation were each to be anchored to a finger of the doctor's left hand, or more precisely to the kinaesthetic representation of a finger. In evoking the doctor's own body image as a part of an almost meditative strategy of awareness-raising during the process of consulting, this account invited a new presence into the consulting room in the form of a body which was not that of the patient. This body was at once useful as a mechanism of non-verbal perception and potentially dangerous as the source of unconstructive motivation and feelings of desire or frustration. Thus Neighbour described a 'red-light zone', an area in which the doctor's own needs for 'physical and psychological homeostasis' were to be kept safely separated from other areas that had a more positive contribution to make to the consultation. Feelings, which enabled the doctor to be intuitive and responsive, were to be admitted into the consultation space, while those that prevented her from being rational and altruistic were excluded. The internal space of the doctor thus had a structure, like a house, which required maintenance – as recognized by Neighbour's fifth checkpoint of 'housekeeping'.

A new space had opened up in which the doctor's subjectivity was located and in which pathological processes might occur during the consultation, manifesting themselves as signs and symptoms within the doctor just as they would in a patient. At a time when new models of health behaviour increasingly required patients to carry out as much as possible of the processes of surveillance, regulation and control of their own bodies and minds, doctors too began to be described as spaces with internal sanitary boundaries, the maintenance of which required vigilant and skilful self-management (e.g. Chambers and Maxwell 1996; Smith 1997). Doctors' own illnesses now faced the objectifying procedures of medical research and practice

The Inner Consultation marked the next phase of a developing trend in the examination of the doctor-patient relationship. It had begun in Balint's groups with the construction of subjectivity by means of a dialogue between colleagues. This led inexorably, via the application of objectifying empirical methods, to the image of the self-regulating doctor effectively alone in a room with her brain in which not only the patient but also her own body was represented as data, all of which in its turn was subject to the same techniques of diagnosis and management.

In 1993, a group of Balint collaborators revisited the original project (Balint *et al.* 1993). They were persuaded that consultation techniques had been sufficiently refined over the years to make it now possible to get nearer a 'whole person diagnosis'. Yet despite this record of success they reported that the 'scene is still entirely patient-centred' (Balint *et al.* 1993: 13). The next task was to extend this vision to bring in another figure, long absent from the picture, but vital for its new integrity: 'After bringing the doctor into view our canvas is nearly complete' (Balint *et al.* 1993: 13).

Observer to subject

In the beginning there was a disembodied, disinterested eye, ('the clinical gaze') contemplating a depersonalized, disintegrated corporal space ('the patient's body'). 'Bodies' were objects that were subject to examinations, operations, or symptoms but not to experience, and the individual doctor's view was indistinguishable from that of the collective 'body' of her professional peers. As the web of relationships in which the doctor was located became more diffuse and complex, the doctor's work became complicated by relativism and subjectivity and it became necessary to construct a new entity: a doctor with an individual 'self', containing an instrument – the 'doctor's mind' – to examine the newly formulated 'patient's mind'. This instrument extended behind the dispassionate eye, but not yet as far as a subjective body; it implicated only those aspects of the doctor's subjectivity most directly relevant to the clinical encounter, to be studied only by the most objective means. This triad of entities (patient's body, patient's mind, doctor's mind) strongly implied the existence of a fourth. Whereas the patient's body was the first of these entities to emerge, the doctor's body, emerging with its hidden baggage of self-interest and desire, was the last.

The body of the doctor was a sort of proto-space, into which could be projected the existential pain of patients and the organizational strain of the health care system, providing a location for both intuition and stress and a field upon which the conflict between the personal and the professional could be played out. Its shadow could sometimes be seen in the form of a communicating object, a puppet performing a strange dance on the video monitors of its colleagues like a honeybee on the threshold of the hive. Its presence was indirectly suggested in expanding accounts of the vulnerability of doctors. When their health was threatened by abstract entities like 'mental illness', 'stress', 'substance abuse' and even 'suicide', doctors, unlike patients, had no terrain, no physicality upon which these external agents acted. Their distress could only be defined in terms of their inability to work; yet accounts of doctors' pathology seemed strangely incomplete without an analysis of the effect of this work on their own bodies.

These imperfections tended to be reported in the general practice literature as if they were setbacks, but somewhere in the newly invented spaces between the personal and the professional, between diagnosis and treatment, between the 'feelings' that constituted the doctor's own experience and the 'behaviour' with which she hoped to affect that of the patient, a new entity was being constructed. This would be an object thus far most remarkable in the discourse by its absence; a newly materialized observer; a common point of origin for both medical perception and action. The prism of perception was slowly rotating.

16
The Subject of Knowledge

The doctor's identity as psychological being and physical body began to be consolidated in the closing decades of the twentieth century. In parallel, the great unified eye of medicine began to fracture into a multi-faceted gaze, each component underpinned by a personalities and idiosyncrasies that belied the formal professional status to which doctors had traditionally laid claim. This meant that the vulnerable mind and susceptible body of the individual doctors could be exposed to the view of everyone. It also meant that the transformative process whereby the student was made into a professional could be made transparent, no longer shrouded in initiation mysteries. From *The student physician* of 1957 that stressed how professional status was achieved (Merton *et al.* 1957), through *Boys in white* of 1961 that emphasized the survival value of student culture (Becker *et al.* 1961), to *The Clinical Experience* of 1981 that addressed the transfer of the cognitive framework that underpinned medical observation (Atkinson 1981), the secrets of the great eye were laid bare. With that revelation, visibility was effectively reversed. For nearly two centuries, the eye of medicine had surveyed and fashioned the patient; now the naked form of the doctor was revealed to a penetrating gaze. The subject of knowledge had become the object of knowledge.

Interpretations

In 1966, Foucault published *Les mots et les choses* that was translated and published in English in 1970 as *The order of things* (Foucault

1970). In this text, Foucault provided a history of the main forms of knowledge that had developed in the Western world over the preceding centuries. His account focused on the changing relationship between words and the things they denoted (hence the title of the original French edition): in the earlier period they were one and the same, later their connection became more arbitrary. He explored these changes through the history of a number of different fields of study, including those that attempted an understanding of the natural world. A fundamental change in the latter, he noted, occurred around 1800 when the new science of biology replaced natural history.

In the eighteenth century, the science of natural history involved the observation and classification of objects in the world of nature, from butterflies and trees to cloud formations and geological strata. The systematization of botany by Linnaeus was perhaps one of the most famous of these attempts to order the natural world. Then, when biology replaced natural history, it redrew the map of what was to be included in its field of study. Biology declared the study of life as its basis, a framework that excluded inanimate objects such as minerals; for those objects that remained (plants and animals), biology replaced the former focus on their outward appearances with an emphasis on the inner forces that differentiated them from the non-living world. More significantly, biology included Man within its new field of observation.

The inclusion of Man as an object of study within the domain of nature rendered Man accessible to investigation, for example, through human biology. Yet it was not simply a case of the new biologist placing Man under the microscope or on the dissecting table alongside all other living organisms since this would be to ignore the special status of this new object of study. For it was Man who studied Man and therefore established a new tension between the observer (the subject of knowledge) and the thing that was observed (the object of knowledge).

In the eighteenth century, the subject was the natural historian examining the object of nature-without-Man; in this guise, the scientist was a disembodied observer of nature. The inclusion of Man as an object of study, however, compromised the purity of this subject-object split that had ensured that the status of the knower was firmly separated from that which was known about. With the new biology and its introduction of Man as an object of study, scientists still stood outside of their field of study, scientists still remained disinterested observers, but that separation had become a little more uncertain. In particular, while subject and object could remain distinct for those natural sciences that examined the substantive matter of the old natural history (though in certain more 'abstract' areas such as physics the relationship began to become more of a problem in the mid-twentieth century), for the 'new' sciences of Man, human biology, anthropology, clinical medicine, and the like, the observer began to examine an indirect reflection of themselves, albeit in anatomical form.

The change in Man's investigative status marked the beginning of the long struggle to keep a separation between the observer and the observed that for the next two hundred years was to provide the context for conceptual and methodological developments within the human sciences. From theories of measurement to problems of bias, from the early twentieth century Hawthorne experiments that showed that researchers could influence the behaviour of the researched to the ability of the researcher to influence their subjects in the Milgram studies of the post-war years, the human sciences fought to maintain a position of objectivity in the face of the constant challenge from a blurring of the separation between object and subject that constantly threatened to collapse the distinction.

Then, during the final decades of the twentieth century, the rhetorical and methodological activity that maintained a scientific and 'objective' separation between subject and object began to dissolve. The reversal of the medical investigative eye that for nearly two centuries had gazed on patients and their illnesses meant that the observing doctor became an object/subject as she herself was dissected. This was a revolutionary change. Not only did the new framework disrupt the objectivity of the observer but the whole basis of the relationship between knowledge and the 'knowing subject' began to be transformed. Such revolutions need a symbolic date from which past and future can be plotted and the years around the early 1960s have particular salience. It was then that the dying were first asked to confess, it was then when hospital bed numbers started their precipitous decline, when notions of risk began to spread, when the integrity and purpose of the medical profession came under challenge, when doctors enquired after subjectivity whilst reflecting on their own minds, when the neuroses and risks came to haunt populations. Whereas the mid-nineteenth century holds the key to the construction of the anatomical body of Man, the early 1960s mark the point at which a newly subjective autonomous identity was finally consolidated and the gaze began to turn on the observer. Such a change in medicine can therefore be used as the point of articulation for new ways of reading the late twentieth century and placing the more distant past in a new context.

Historical discontinuities

In 1963, Foucault published *La naissance de la clinic,* (translated and published in English a decade later as *Birth of the Clinic* (Foucault 1973)). This text described the emergence around the turn of the eighteenth century of a new clinical medicine that identified the pathological lesion as lying at the heart of illness and the hospital as the neutral space in which to tend the sick. Here, revealed for the first time, was the birth of the medical paradigm that was to dominate clinical work for the following two centuries. The revelation of Foucault's text made possible a new way of reading medical history (and provided the possibility and framework for the earlier chapters in this book).

Foucault's history of changes in medicine at the end of the eighteenth and early nineteenth centuries belonged, like his 1966 study of the sciences of Man, to the moment of transition in the twentieth century, when the field of knowledge and status of the observer were transformed so releasing the possibility of a new interpretation of the past. As previous chapters of this book have described, the 1960s witnessed a number of fundamental transformations in medicine. Therefore, *Birth of the Clinic* at once stands in a context that made its writing possible, and in a reciprocal gesture, made possible a particular history of medicine. Knowledge of the medical past only emerged in 1963 and it is only from this vantage point that the years before 1963 were comprehended anew, ordered and made into a coherent pattern.

Before 1963 there were histories of the rise of medical science but these were largely progressivist accounts of the triumph of clinical acumen, linear pedigrees for the great successes of medicine, of the unfolding of knowledge as truth was wrested from darkness, at best histories of medicine as an institution but not as a cognitive system. After 1963 a disjunction appeared, separating past and present by a new chasm and, simultaneously, establishing the truth of a massive disruption in medicine two centuries earlier.

The principle of reversing visibility, of the rotating prism of observation, was not confined to medicine. Contemporaneously, another set of formerly disembodied and anonymous investigators began to emerge. In *Laboratory Life*, published in 1979, Latour and Woolgar described the 'non-scientific' behaviours of scientists when engaged ostensibly on scientific tasks and experiments (Latour and Woolgar 1979). In place of working lives being governed by a dispassionate objectivity as scientists sought to reveal the truth of nature, Latour and Woolgar found that they indulged in various subterfuges to ensure success. Indeed, their behaviour could be construed as being more motivated by personal vanity than by the search for objective knowledge; their approach was far removed from the claim to disinterested observation by which they justified their position. In other words, a close examination of the everyday routines of scientists, revealed their work to be governed more by subjectivity than objectivity: the latter was simply the scientists' own rhetoric for the public domain.

Observations of scientists at work began to open up the mind of the scientist to public scrutiny. Just like the doctors – and at almost exactly the same time – the disembodied eye of the investigator began to develop a mind, a psychological presence, that before had been constituted by a generic and formal identity of 'scientist'. This mind was not standardized, a uniform product of a common educational process and professional mission, but showed great variability. From scientists engaged in fraud, to scientists jealous of colleagues, from scientists working hard, to scientists cutting corners, the range was analogous to those individual differences uncovered in every-Man by these same anonymous observers early in the twentieth century. The mind of the observer was thereby being constructed in the reflexive gaze of scientist observing scientist in a great web of reciprocal revelation, just as doctor was beginning to construct doctor. Professional doctors; disinterested scientists; those great totemic images of a distant observing class, of a fundamental cleavage between observer and observed, of subject and object of knowledge, began its process of disintegration.

In 1998, Lawrence and Shapin published a collection of essays on the bodies of famous scientists as a long neglected aspect of scientists' lives. These were not the bodies of anonymous stereotypes but of named individuals providing their own bodily reports, suffering their ailments, enduring their feelings. Rather than the conventional hagiographic account of scientists wrestling with great thoughts, the picture was one of ordinary bodies experiencing mundane symptoms and illnesses. Nevertheless, this transubstantiation of the scientist's body was less a discovery of an ignored fragment of history and more the final embodiment of the observer. At exactly the same time as surveys of doctors were discovering stress and illness, Lawrence and Shapin made their claim to have 'discovered' a new perspective on the lives of scientists. Surely, scientists were no different from doctors in this regard – flawed, embodied, constituted – just like other bodies? Yet it had taken nearly two centuries for the figure of the observer/investigator to emerge from behind the prism of visibility.

In 1968, Barthes first published his celebrated essay, 'Death of the author' in which he pronounced the end of the author as the central figure of literature (Barthes 1968). His analysis revolved around the new centrality of text. Text consisted of words written by an author and read by a reader. Text had therefore simply seemed the medium of communication between author and reader; the author's thoughts were paramount and, in the absence of a verbal performance, the text allowed their dissemination to a wider audience. This framework privileged the author as the point of origin of the text, as the figure at the centre of literature.

But what if the text could be seen as having a measure of independence? If the text was divorced from the author and the act of reading was not simply the indirect transfer of the author's mind into the reader's but involved an active process of interpretation, then the connection between author and text would be severely weakened, if not lost. If a text needed active interpretation, then the resulting inferences might differ from reader to reader: which reader had abstracted the author's original thoughts? Who could know? Even the author's own account would itself be a retrospective interpretation of his or her own text. There was no truth contained in the text, only a multiplicity of interpretations.

In many ways, the relationship between author and text mirrored that between investigator and the object of their investigation. The author/investigator had been engaged with the production of a 'truth' that lay outside of themselves and as such they had been disembodied narrators whose identities could only be inferred from their texts or status. In the 1960s and 1970s, however, this dispassionate and separate relationship began to crumble. Scientists were in some way a part of the objects they studied; the doctor was some sort of reflection of the patient; the author was an artefact of particular reading of a text.

In sum, between the late 1960s and the end of the twentieth century the disembodied gazes that for two centuries had been such an important part of the analytic fabric of knowledge began to materialize from behind the figure of the anonymous observer. In the 1960s, the problem of authorship was projected onto the stage of textual communication at the same time as the doctor's identity as an artefact was being explored in studies of medical socialization. Then, as doctors idiosyncratic selves began to emerge in the 1970s, the figure of the scientist entered the same deconstructive framework as studies showed that individual 'styles' of scientists existed in much the same way as the individual styles of doctors. And by the final decade of the century, as the doctor's physical presence began to crystallize, the scientist's body also made its first appearance, not the formalized or stereotyped body of some 'ideal' or caricature but the ordinary vulnerable body of every observer.

The stylizing of authorship fragmented its once unitary gaze. Every doctor, every scientist had her own repertoire of practice. Further, these repertoires were of relevance to what they 'saw': the object of observation was in some way an artefact, a result even, of the idiosyncratic style of the observer. Methodologically this created problems for the supposed 'objectivity' of science/medicine and new techniques were needed to accommodate the fragmenting reports of the great eye of observation. Such methods were not rooted in their ability to establish some correspondence with a world-out-there but in their capacity to harness disparate subjectivities. One new repertoire of techniques was consensus methods that offered the possibility of rebuilding a gaze that was rapidly losing its once formidable unity.

Consensus methods as formal techniques for establishing agreement were developed by psychologists in the 1950s and spread to other disciplines, including medicine, over the next two decades. This new approach to investigating the world-out-there was formalized in three different methods. The Delphi technique, which involved an iterative questionnaire going between group members, was described in 1963 (Dalkey and Helmar 1963); this was followed by the reporting of the nominal group technique in the late 1960s (Delbecq and Van de Ven 1971); and finally, in 1977, the National Institute of Health in the US introduced the consensus development conference (Fink *et al.* 1984). This formalization of the process of collating subjective judgements also found expression in statistical techniques and approaches, from Cohen's kappa in 1968 that summarized degree of agreement (Cohen 1968) to the challenge of Bayesian statistics that attempted to revitalize statistical inference through the incorporation of the scientist's subjectivity. In a world that was beginning to conflate the objective and the subjective, the urgent task for science and medicine was to develop techniques that could convert subjectivity into objectivity and so maintain their straining separation.

Yet, even as they sought to maintain an ordered world of objectivity, the new methodological techniques could not conceal or overcome the existence of subjectivity in the former exclusive field of objective knowledge. The subjectification and materialization of the observer meant that the integrity of knowledge that had been produced and maintained through a disembodied analysis began to dissolve and with it a new space of enquiry, a new plane of knowledge opened up. This analysis prioritized the text (as narrative): the text reflected neither objective truth nor the machinations of the author's mind – it existed between the two, in a new space divorced from its past immutable referents.

This new plane of knowledge began to emerge in the 1960s and by the turn of the century was well established. Old certainties disappeared; a linguistic 'turn' afflicted knowledge; a post-modern world destroyed the meta-narratives of the past to replace them with fragmented, ephemeral and localized knowledges. The new plane of knowledge also opened up the possibility of a reinterpretation of history – and other creation stories.

Readings and authorship

The Biblical creation story and its accompanying commentaries had allowed nature to be read as a book of praise to the creator: everywhere God's work was bountiful and illustrative of his omnipotent power. When Darwin outlined his revisionist account of origins, he also presented a new language for reading the evidence of long-extinct organisms and progenitors of Man. The new Darwinian language allowed each of the elements in the book of nature to be read in a new way, as evidence both for the truth of the new language and the veracity of the new way of seeing. A fossilized skull fragment chipped out of a rock bed became corroborative evidence of Man's origins and, in a tautological movement, further support for the underlying theory that allowed an otherwise nondescript artefact to be read in this way.

Darwin's theory, however, is more than an explanation of origins. It is also itself a text that can be read in different ways. For over a century, there has been a unified reading (claiming, of course, to reflect Darwin-the-author's original thoughts); this reading – as befits a creation story – pronounced on the truth of origins. A different reading, as described in previous chapters, tells a different story, one that situates Darwin and the Darwinists as a part of the very process they claimed to have revealed. Thus, Darwin and Darwinist writings, together with other texts on Man, from their nineteenth century beginnings to their late twentieth century consolidation can be read as describing the recent emergence of Man as an object of study, as a very recent construction of words and practices. From the delineation and segmentation of a distinct corporal space in the mid nineteenth century, through the realization of movement and subjectivity, to the actualization of reflexivity, the history of Man can be read from medical texts, from clinical practices and from hospital architecture, from laboratory tests and from health care organization. Each register provides a parallel reading corroborating and fleshing out the new creation story of recent origins. Where Darwin presented a reading of nature to locate the origins of Man, this story advances a reading of Man.

Yet there is a further and more fundamental difference between Darwin's creation story and the account of recent origins presented here. Darwin the observer – to whom can be added Darwinists – existed apart from the phenomenon that he and they wished to study: palaeontologists might have traced their origins to the rocks they deciphered but they themselves were separate beings from the fossils they studied. There was an asymmetry between rock and observer: the former was the object of study, inert, unprotesting, waiting to be read, while the latter was the subject, active, aware, interpreting, the reader of the text. This separation, however, became less tenable in the closing decades of the twentieth century. For then the texts that were interpreted – books, bodies, architecture, practices – became themselves the active product of Man 'the subject'. It was 'Henry Gray' who wrote the great textbook on human anatomy, describing not only the bodies he dissected, but also his own. Reader and text were not dissociated within the prism of observation; it was not nature that laid down the evidence for Man's origins but Man himself.

The late twentieth century recasting of the optics of observation allows an answer to the question of the relationship between text and Man: how can Man invent Man, how can identity invent identity? This book has used texts, notionally written by Man (in the form of authors), to write a new creation story of Man's origins. How could authors exist before their own invention? The answer hangs on the primacy of text and on the construction of authorship. The answer involves a re-examination of the status of the subject, of the observing eye. Instead of prioritizing authorship, emphasis on the text allows the author to become an artefact of the text: the 'real' Henry Gray comes to exist through his text. Likewise, the identity of Man is a creation of the words and practices used to grasp at the inchoate world glimpsed through the prism of observation.

The changes in the early 1960s that reorganized the field of knowledge might suggest that the human science project of the preceding century and a half to keep subject and object separate was coming to a close. The dream of veridical perception, of true knowledge, was ending. Perhaps the triumph of the human genome and all its promise for transformation mark the over-blown dream of joining a myth of origins to an imaginary future. Perhaps the New Darwinists with their subjugation of the socio- to the bio-, their selfish genes, and their behavioural genetics sensed that the old Darwinian story needed up-dating for a new epoch as they searched anew for the solid referents that could define Man's true identity. Yet restoring the walls of the Darwinian citadel does not address the form and direction of the looming threat.

The new medical revolution, the new creation story, the prismatic reading of Man, all depend on certain conditions of possibility that only appeared about 1960. As previous chapters described, this period witnessed the rotation of the prism of illness to focus on the doctor, contemporaneously with the reversal of visibility between scientist and 'nature' and between author and text. It is as if time became concatenated. *Birth of the Clinic* documenting the triumph of a medical system at the moment of its demise; the transformation of death in the confession of the dying as the majesty of pathological death was revealed; a new appreciation of the dominance of the hospital as its suzerainty started to crumble. Was there a history before 1960? Or were a new past and a new future constructed in tandem, an Archimedean point from which time itself was invented through the seismic shifts of the very recent past.

17
A Note on Methodology

Preceding chapters of this book have provided a history of Man, an account of how a corporal and psychological identity was forged over a period of less than 200 years. The method used has been 'historical', but this is not to imply that such an approach has any greater constancy than the objects it describes. Indeed, the method of the book is itself intimately bound up with the picture it portrays; history itself is part of the story. Yet rather than weave another thread, this final chapter – more of an appendix than a continuation of the previous narrative – tries to make transparent the approach that has been adopted and the tensions that result.

History provides an account of the past. Before the eighteenth century, this account would seem to owe as much to myth and storytelling than to what actually happened. Then modern history appeared with its disinterested search for an 'objective' narrative of the past. In a sense, this rigorous search for the truth of the past became so clearly the mark of the historian that it hardly needed methodological texts to describe and justify. What it needed were exemplars, illustrations of how it strove to build better and better representations of the past. Of course, there were those who failed through either dishonesty or an ideological commitment to a particular and distorted reading of the past but they hardly merited the title of historian; and besides subsequent histories could offer suitably corrective interpretations.

In 1961, Carr published his celebrated volume *What is history?* (Carr 1961). He discussed how singular events became transformed into a narrative 'history', suggesting the possibility of a more creative role for historian than simple reportage, but in the end concluded that history was possible and that commitment to an open mind and rigour would enable truths about the past to be revealed to the present.

The significance of Carr's reflection on history's purpose lies less in what he said and more on when he said it. It was the second half of the twentieth century in which questions began to emerge that challenged the foundations of the historian's craft, particularly in respect of the nature of history and the application of historical method. For Carr, an 'objective' history was still ultimately defensible; yet others began to exercise a self-reflective awareness of the assumptions behind their work that culminated in the view that history was more 'invented' than 'discovered'. In other words, accepting that there were billions of 'events' in a historical past, the historian added an interpretive framework as a means of organizing a very few selected events into a pattern or story. In this way, there could be many histories of the past, each dependent on the choice of events and the interpretive gloss used to establish a coherent narrative. The historian was therefore closer to the novelist in constructing a good story than the image of the scientist supposedly revealing aspects of an objective world.

In this light, the history described in previous chapters can be seen to involve events or facts taken from medical texts and woven into a story of origins. This story is only one that could emerge from these texts but its meaning is based on the final goal, accounting for identity. In this sense, it is what Foucault would describe as a history of the present rather than of the past, telling not what went on before but how we arrived in the 'now'. With this purpose, the method deployed in the text had to address two significant conventions of historiography, namely the distinction between primary and secondary sources and the place of the author in relation to the text.

A hierarchy of texts

Students of history are taught the difference between primary and secondary sources. This distinction is similar to the legal difference between the actual witness to an event and the second-hand report or 'hearsay' of someone not actually present. In the same way, a model is presented in which historians study 'original' source materials such as diaries, registers, letters, or chronicles then subject them to scrutiny, analysis, interpretation and synthesis. The result of this creative endeavour is a 'secondary source', a new rendering of the historical record informed by a retrospective examination of previous texts. Secondary sources, either devised by someone removed from the original events, or another historian could then be used, in their turn, as the basis of further historical accounts. For the historian, however, the possible editing and refraction that took place between past events and their later description meant that a higher level of critical suspicion has to be employed for secondary sources.

In many ways, the distinction between primary and secondary sources is an ideal type, of particular value in pedagogy, but not necessarily of use to the historian who must view all sources with a critical eye. Even so, the distinction between primary and secondary sources embeds an important concept to which many historians would subscribe, namely the separation of the interpretation of the evidence from the evidence itself, a separation that implies that observers can be held apart from the historical events external to them. The primary source or original text represents the immutable bedrock that can always be visited again and again, retaining its constancy while yielding different subjective interpretations. Equally, a secondary source or text itself can be used to create further secondary sources each one a point further removed from the original events. In this way, a hierarchy of texts is established ranging from the primary untainted evidence to the interpretations of interpretations of layers of secondary texts.

The closing decades of the twentieth century, however, witnessed the 'death of the author' and the new prioritization of text (Chapter 16). The distinction between primary and secondary sources and texts then became somewhat arbitrary. Instead of the authorial presence distilling primary knowledge into a secondary rendition, the text invited the reader to construct the author and her imagination. Neither primary nor secondary text had priority in this interpretive process as all texts were interpretations; one text could not be more 'interpretive' than another. Text, author/historian and time were all constructed at the moment of reading: a reading in 1930 would offer a different view of the past from a reading of the same text in 1990. In other words, a primary source may notionally have been an account of some past events, but these events had themselves been selected, interpreted, edited and collated to produce a coherent narrative; it was only raw or primary to the extent that professional historians or other later commentators had not fashioned it.

The problem can be posed in terms of whether to stabilize the present or the past. The source is a text (mostly a written text but at least something that is 'read'). Texts need readers to bring them to life and establish their core messages. The act of reading in the present produces three possible histories: one is history as a story about the past; another is a story as constructed by a historian in that historian's present; and the third is an immediate history constructed at the moment of reading. Alternatively, by stabilizing the past we can construct a fourfold classification of historical material. First, there is the 'pre-text' moment of the event, or rather events as many things happen contemporaneously in time. The pre-text moment is surely 'raw', filled with myriads of trivia, of actions, of states, of movements. Then there is the primary text, the conventional primary source, an account that selects out some of those infinite series of events at the pre-text moment and makes sense of them as in a register. The secondary text (and there might be layers of them) is the secondary source, a synthesis by the historian, or others, of primary texts, a distillation informed by some purpose or agenda, a contemporary gloss on times past. Finally, there is the present or point of reading.

The pre-text moment remains inaccessible; it is impossible to encapsulate everything about the present never mind a moment or a century ago. The primary text is the later selection and preliminary editing of the rich diversity and chaos of the pre-text instant: as such, it must tell as much about the moment of its creation as about its supposed object. The secondary text must also reflect the values, mores and views of a yet later time. In other words, the same 'events', like ripples on a pond, spawn resonating waves of interpretation that can reveal as much about their own moment of creation and how the past is constructed than about the past itself. Any primary or secondary text, therefore, embodies the encrustation of later times on the supposed pre-textual events that they notionally record. Of course, that original recording of pre-textual events, that initial splash in the pool of time, constrains subsequent secondary interpretations thus giving a sense of continuity, of historical linearity, to the layers of text. Then the reader who reads with an immediate prism of perception brings the time of creation of the primary and secondary texts into the present.

For example, in any year of the eighteenth century there were many events. Some of these occurred in a world defined by medicine – patients in pain, patients eating, sleeping, defaecating, patients gasping a final breath, a doctor walking, attending, listening. Pomme, an eighteenth century physician, chose to describe one such series of events in terms of what he saw discarded in the urine of one of his patients. Pomme is, then, the primary source, the first distillation of pre-textual events into text. In 1963, Foucault opened his book *Naissance de la clinic* with an edited account of Pomme's observations and then went on to juxtapose this with a later nineteenth century description of the layers of the brain's covering (Foucault 1973). Between these two descriptions, he argued, lay a perceptual and epistemological chasm as medicine was revolutionized by the new clinical practice based on the underlying pathological lesion. Foucault, the historian, therefore was offering a secondary source, an interpretation at a later time. But does it tell us about its historical subject (some events in the eighteenth century)? Or Pomme's perception of those events? Or about 1963? Or about now?

The previous chapters have tried to flatten the hierarchy of texts. This means that a book written about the eighteenth or nineteenth centuries written in the twentieth belongs in the narrative to the latter. So, Foucault's description of the late eighteenth century 'Birth of the Clinic' was determinedly placed in 1963 (its year of publication in French), to an age that began to periodize its history of development. From Kuhn's 1962 novel description of science as existing in constraining paradigms (Kuhn 1962) to Foucault's recognition of changing epistemes (Foucault 1970), histories of the past were constructed that told more about their 'present' in those important years at the beginning of the 1960s. It was as if Foucault could only describe a history of the modern medical paradigm at the moment when the old order was coming to a close and a new perspective was being born. It was as if the birth of knowledge about the past could only be realized at its moment of its demise. In this sense, Foucault's history belongs firmly to 1963 as it could not have been written before; and the years prior to 1963 become a construction, an artefact, of that year and its particular reading.

In sum, a text can be read in two ways: as a description of a past it professes to record (the late eighteenth century in Foucault's case), or as a refraction of the time it was written/published (1963 in this example). For the purposes of this book it is the latter model of history that is deployed in that all texts are treated as primary, as constructing the world at the time of their publication. Secondary sources, histories, commentaries, past reflections are all aligned with their date of publication.

It follows that this book can make no attempt to question whether any text is 'true', that is whether it accurately represented pre-textual events. How could such a question be answered? It is not as if the 'rhetoric' of how, say, doctor–patient interaction was being conducted can be checked against an independent empirical world. The latter, if it existed, is itself produced as text, yet another truth-claim to be juxtaposed against others. How then to distinguish one claim from another, to select the more true text? By consensus? Then truth is established by weighing texts. By (theoretical) coherence? Then truth is established *a priori* in the framework imposed on textual interpretation. No, the only approach is to accept that all texts speak the truth, that texts do not lie. They should not be rejected by juxtaposition against some arbitrary external referent or because of their ideological or political position. Each text belongs to a regime of truth and the task is not to judge that truth but to make clear its relationship to the present. When, in the eighteenth century, Pomme saw the insides of a patient discarded in urine ('membranous tissues like pieces of damp parchment ... peel away with some slight discomfort, and these were passed daily with the urine') Foucault assured us that the account should be taken seriously ('How can we be sure that an eighteenth century doctor did not see what he saw ... ?' (Foucault 1973: x)). Equally, when texts that claimed to encompass all mortality made no mention of infant deaths it is reasonable to assume that they had no independent existence. Or when a text of the 1950s on clinical examination made no mention of male sex organs it is reasonable to accept that, at that time, for that regime of truth, they did not exist.

No doubt part of the problem is the Enlightenment language through which history constructs itself. The past is 'discovered' or 'revealed' or 'found', implying that it previously existed but was hidden. Far better, surely, is to use the word 'invent'. Claims such as the neuroses were invented, heart disease was invented, or social support was invented all throw into the kaleidoscope of interpretation what previously seemed so real. In other words, anything can become real, and then can disappear. Humours existed for eighteenth century medicine, but not now; cells exist for twenty-first century medicine (but for how much longer?); meridians apparently exist for acupuncturists though they cannot be seen or measured with conventional scientific equipment. We re-write our histories with alarming arrogance as we air-brush 'facts' from the historical record. All those deaths from fatty heart in the middle of the twentieth century that in retrospect are pronounced mis-diagnoses; all those instincts guiding conduct that were usurped by more contingent psychological constructs like 'attitudes'; or the millions of men who (briefly) had Type A personalities in the late twentieth century before they were endowed with something else more suited to the age (Riska 2000).

In the same way, the understanding of 'causes' of events lies within specific regimes of truth. An acceptable explanation or perceived cause of a phenomenon at one time will not necessarily be adequate at a later date. This makes it difficult to offer a causal explanation of how one regime of truth changed into another, how one way of understanding identity mutated into a new one, as there is no explanatory system that transcends the change in explanatory systems that is being explored. The problem applies in particular to explanations of origins, to what 'caused' God or what 'caused' evolution. Ultimately there is only a functionalist answer to these fundamental questions, of explaining the phenomenon in terms of what it produced rather than what were its antecedents.

One implication of the approach to texts and causal explanation adopted here is that the intellectual antecedents of this book are not explicit. The texts and ideas that 'influenced' me, the author, are absent because this juxtaposition of texts would establish yet another layer on history. Unfortunately, this does a disservice to the many authors to whom I am indebted; it also means that my own antecedent writings are hidden from view. It is not that they cannot be mentioned only that they can be analysed except in a context that situates them in the time they were written, otherwise they would threaten a return to a hierarchy of texts.

Another implication of thinking about historical texts in this way is that we can reorganize our libraries. Why are books classified by subject and/or author? Why is it that I can find a section of the library on 'crime novels' and another on 'books about nineteenth century medicine' and in these sections I can find them grouped by author? The alternative system of classification is to reorder all books by date of publication: books on nineteenth century medicine jostling with crime novels and books describing the seventeenth century. What would be apparent are the threads that hold such disparate texts together, the contemporary framework that underpinned all texts written at the same time. Laennec's account of the invention of the stethoscope of 1819 (which offered a method for analysing bodies) sitting comfortably next to Shelley's novel, Frankenstein; the literature on the placebo (on how the body is influenced through the mind) beginning in 1948 set against the McCarthyist texts on the communist threat or the dangers of brainwashing in Korea or Orwell's *1984*.

In effect, the convention of classifying texts in libraries according to their author (especially for 'fiction') and by subject (for 'non-fiction') conceals the centrality of temporal ordering and in so doing, ironically, sustains the mirage of a hierarchy of texts. Classification rules that have been applied to texts – authorship, subject matter, primary or secondary – can all be seen as devices to maintain the separation of subject and object, of author/historian and immutable fact, of the interpretive act from the domain of truth, across a linear time-line. What a disturbing world lies beyond a simple reordering of books in the library.

A non-hierarchical reading of texts can establish a development trail like fossil layers in sedimentary rock. Different editions of the same book, for example, can be used to trace the emergence of new objects. Between 1935 when Hutchison's *Clinical Method* suggested asking the patient 'What is your complaint?' and 1976 when it had changed to 'Now please tell me your trouble' lies a change in the very nature of who we are. Equally, when Muir's *Textbook of Pathology* of 1933 claimed a bacillus caused tuberculosis it represented a very different model of the world from his later claim in 1951 that the cause was multifactorial. Who knows what 'really' causes tuberculosis (as if the question could be finally answered) when contemporary explanatory frameworks provide the structure for the regime of truth that determines the crucial distinction between what is real and what is not real, between truth and error.

The strategy is not to play the common historian's game of identifying earlier and earlier examples of primary sources as that only serves to structure the continuity and linearity of historical development. The 'discovery' by one secondary text of a primary serves as a challenge to reread 'earlier' primary texts for even earlier examples. For example, it was is only when a 'secondary' text first identified and explored the idea of patients' 'emotions' that the chase could begin to find such expressions in primary texts – in diaries, in letters, in case reports, in re-translations, etc. – from an earlier and earlier date, even though, alas, its origins lie only in the secondary text read as primary. Did patients have heart attacks before the first identification of heart attacks in 1912? Yes, but only by an analysis that organized past events in a new post-1912 synthesis. Did patient's in the past have a 'view' of their illness? Yes, but only in retrospect from the early 1980s when it came into historical consciousness. Did Leonardo or Vesalius identify the 'anatomical body' of everyone earlier than the mid nineteenth century? Yes, but only since the mid nineteenth century.

The act of 'reading', of course, holds all these layered accounts together by locating the text in the year in which it is read. The description here – all the above words – relates to a reading at the opening of the twenty-first century; no doubt sometime in the future it will be possible to offer a different reading based on the new vantage point of contemporary perceptions and concerns. This process of reading is the means through which history is actively produced as an on-going enterprise; it is also the process through which a temporal past is refashioned, reformed and extruded as a familiar linear dimension.

The reader provides the over-arching view, constantly reinterpreting, however subtly, constantly reinventing, achieving a congruence and stability between all those previous readings. Therefore, only when these readings began to dissociate in the closing decades of the twentieth century, their differences began to become apparent. It was the realignment of text and author; it was the growth of a hermeneutic tradition in the twentieth century that made meaning problematic; it was the emergence of 'interpretative' explanatory frameworks. The result was that the solidity of historiography began to crumble. Perhaps Carr's *What Is History?* of 1961 might have unwittingly opened Pandora's box? More likely, he noticed that the box was ajar and tried to close it. But he failed for others engaged with new ways of thinking about the past. Their subsequent texts belong to that fertile period in the early 1960s that turned in on itself to reveal the layers of reflection and refraction that might be reconstituted for a brief moment by the reader.

History without an agent

In his essay *What Is an Author?* Foucault identified a shift in the 'author function' over the centuries (Foucault 1977). In Mediaeval times, the truth of the text was to be discovered in the truth of the author – a saint, being who he was, only spoke the truth while heretics were known to write untruths. The image of the book-wheel of the learned monastery typified the analysis: a series of books laid out on a circular rim that the author rotated while adding paragraphs and sentences from existing text to create a new synthesis containing the truth of previous distillations of knowledge. In modern times, however, that relationship has been reversed; the truth of the author is to be found in the truth of the text as we scan the author's words to find out who he or she really was. What was the author really like? It depends on what they said in their text. Nursing textbooks prior to about 1970 offered a way of managing the dying derived from Nightingale's own *Notes on Nursing* (Nightingale 1859). This involved not telling the patient they were dying, offering physical comfort and remaining cheerful at all times. After 1970, nursing textbooks changed their advice. Patients should be told and counselled while the nurse herself should reflect on her own death. And the rationale for this position? Nightingale's own words in *Notes on Nursing*. So who was Nightingale? Before 1970 and after she was two different people, each a reflection of the particular reading of her text.

This understanding of the relationship between text and reader is important for a history that tries to decentre the author. History places the author or subject at the centre of the world in two ways. First, as described above, the privileging of various texts places the authorial imagination at centre-stage: the author is ever-present as recorder, interpreter or distorter of events. Second, history is about humans, it celebrates a Man-made world (the alternative of natural history belongs to a very different system of knowledge).

In short, for historians, the person is the primary input into the story. It is the person who thinks and the person who acts. If a text wishes to explain 'the person', however, then the traditional polarity must be reversed. If the task is to write a history that removes Man from the input side of the equation and asks instead 'what makes Man?' then a different sort of historical method needs to be used. What this means in practice is that the traditional historical interpretation of texts (Who wrote this text? What was their motive? What was their view? What was their ideological position? etc.) can be rejected and replaced by an analysis that reads text as a mechanism for constructing Man (What sort of Man could this text 'see'?). Man must become the dependent variable rather than the independent. The constancy of Man must disappear.

The problem can be expressed in terms of agency. Agency presupposes a thinking, acting individual who has thoughts and carries out deeds. Yet agency is itself an historical emergent; as Chapter 8 described, agency comes to inhabit the body relatively late in the twentieth century. Therefore, to provide an account that tries to explain agency rather than assuming it, the relationship between thinker and thought, between doer and deed, needs to be reversed. It is the thought that constructs the thinker and the deed that constructs the doer.

This gesture allows the universal model of Man to be overturned, a model that claimed if a human characteristic was absent in any investigation then it had been repressed. Instead the model is of an identity recently created, of absent characteristics that have yet to be forged. The human sciences, for example, in exploring who Man really is, study the characteristics and forms of the infant. History, in alliance, asks the question when was the infant 'discovered', when did this object first come to social attention. The alternative is to view infancy not as a hidden thing lying undisturbed until its moment of discovery (then, in retrospect, identifying all those 'clues' to its existence before its formal revelation), but as an invention, as a product of ways of speaking, of analytic dimensions, of lines of investigation; an object that becomes 'real' for the time when it is consolidated by analytic forces, and then when the gaze turns away, begins to crumble. The question is not when Man was discovered but when Man began. Darwin conflated the two questions: his discovery of early Man was also the story of origins. He also missed the significance of his own contribution to the construction of an object that he thought had a history of millennia when it had a history of only a few years. The moment when it was first possible to draw a boundary around the (ordinary) body of a separate everyone brought the figure of Man into existence.

In this analysis, then, the author of a text is unimportant. Their biography is a siren call that will only lead away from the central task of understanding Man by interpolating a ready-made interpretation. Thus, as described in Chapter 7, Man can be traced as appearing through texts on clinical method without wondering who the author of the text 'really' was, of their background, or biography, or agenda. For this reason, a reflexive history needs to concentrate on 'technical' texts. If individual identity is what needs to be explained there is too much distraction in texts written by authors whose biography suffuses the text or in texts that have preexisting identity as their core assumption. Ironically, it is the routine, almost anonymous, texts such as those on cross-infections in hospital, in blanket washing, in enforced bed rest, on floor disinfection, in which the 'person' is reduced to a peripheral object, that can speak loudest on the nature and changing nature of identity.

From this dissolution of the ordering of texts and the repolarization of agency and authorship, a new figure of identity emerges. The present provides the Archimedean point from which to observe the rotating galaxy of bodies, identities, authors and texts, to confront the paradigmatic Darwinian fable and to construct a history of identity that has its origins less than 200 years ago and began anew in the last four decades.

References

Abrahams, A. (1930) *Exercise: Its Functions, Varieties and Applications*. London: Heinemann.

Aitken-Swann, J. (1959) 'Nursing the late cancer patient at home: the family's impressions', *Practitioner*, 183: 64–9.

Aldrich, C.K. (1963) 'The dying patient's grief', *Journal of the American Medical Association*, 184: 329–1.

Alvarez, W.C. (1952) 'Care of the dying', *Journal of the American Medical Association*, 150: 86–91.

Anderson, K. Coulter, J. and Looke, E. (1960) 'Transfer of staphlococcus pyogenes from infected to non-infected hospital beds', *British Medical Journal*, i, 1925–7.

Apple, D. (1960) 'How laymen define illness', *Journal of Health and Human Behaviour*, 1: 219–25.

Aries, P. (1981) *The Hour of Our Death*. London: Allen Lane.

Armstrong, T.G. (1946) 'The use of reassurance', *Lancet* 2: 480–2.

Arnold, M. and Ware, J. (1953) 'The General Practitioner's premises', *Practitioner* 170: 582–91.

Aronson, G.J. (1959) 'Treatment of the dying person'. In: Feifel, H. (ed.), *The Meaning of Death*. New York: McGraw-Hill, pp 251–8.

Asher, R.A.J. (1947) 'The dangers of going to bed', *British Medical Journal*, ii, 967–8.

Asher, R.A.J. (1951) 'Munchausen's syndrome', *Lancet*, i, 339–41.

Asher, R. (1955) 'Management of advanced cancer', *Proceedings of the Royal Society of Medicine*, 48: 376–7.

Ashton, J. and Seymour, H. (1988) *The New Public Health*. Milton Keynes: Open University Press.

Ashworth, H.W. (1963) 'An experiment in presymptomatic diagnosis', *Journal of the College of General Practitioners*, 6: 71–3.

Atkins, H.J.B. (1942) *After-treatment*. Oxford: Blackwell.

Atkins, J.B. (1904) *National Physical Training: an Open Debate*. London: Ibster.

Atkinson, P. (1981) *The Clinical Experience*. Farnborough: Gower.

Aylett, M.H. (1976) 'Seeing the same doctor', *Journal of the College of General Practitioners*, 26: 47–52.

Balint, M. (1955) 'The doctor, his patient and the illness', *Lancet*, i, 683–8.

Balint, M. (1964) *The Doctor, His Patient and the Illness*. London: Pitman.

Balint, M. (1956) *The Doctor, His Patient and the Illness*. London: Pitman.

Balint, E. and Norell, JS. (Eds) (1973) *Six Minutes for the Patient: Interactions in General Practice Consultations*. London: Tavistock.

Balint, E., Courtenay, M., Elder, A., Hull, S. and Julian, P. (1993) *The Doctor, the Patient and the Group: Balint Revisited*. London: Routledge.

Balint, M., Hunt, J., Joyce, D., Marinker, M. and Woodcock, J. (1970) *Treatment or Diagnosis: a Study of Repeat Prescriptions in General Practice*. London: Tavistock.
Barnard, B. (1952) 'Sterilising blankets', *British Medical Journal*, i, 21.
Barron, N. (1916) 'Physical training, with special reference to the training of convalescents', *Journal of the Royal Army Medical Corps*, 27: 460–75.
Barthes, R. (1968) 'Death of the author', *Manteia V*. Translated and reprinted as 'The death of the author' in Heath, S. (ed.) (1977) *Image – Music – Text*. London: Fontana.
Bascombe, E. (1851) *A History of Epidemic Pestilences, from the Earliest Ages*. London: Churchill.
Batten, LW. (1961) 'The medical adviser', *Journal of the College of General Practitioners*, 4: 5–18.
Baumann, B. (1961) 'Diversities in conception of health and physical fitness', *Journal of Health and Human Behaviour*, 2: 39–46.
Beauchamp, T.L. and Childress, J.F. (1979) *Principles of Biomedical Ethics*. New York: Oxford University Press.
Becker, H.S. (1963) *Outsider: Studies in the Sociology of Deviance*. New York: Free Press.
Becker, HS. Geer, B. Hughes, E.C. and Strauss, A.L. (1961) *Boys in White: Student Culture in Medical School*. Chicago: Chicago University Press.
Becker, H.S. (ed.) (1964) *The Other Side*. New York: Free Press.
Benton, T. (1993) *Natural Relations*. London: Verso.
Berger, J. (1967) *A Fortunate Man*. London: Allen Lane.
Bergey, DH. (1904) *The Principles of Hygiene*. Philadelphia: Saunders, 2nd edn.
Blackwell, B. (1962) 'Hospital addiction' (letter), *British Medical Journal*, ii, 1060–1.
Board of Education. (1909) *Suggestions for the Consideration of Teachers*. London: HMSO.
Bomford, R.R., mason, S. and Swash, M. (1975) *Hutchison's Clinical Methods*. London: Churchill Livingstone, 16th edn.
Bourne, G. (1931) *An Introduction to Medical History and Case-taking*. Edinburgh: Livingstone.
Brackenbury, H.B. (1935) *Patient and doctor*. London: Hodder and Stoughton.
British Medical Association Medical Care Planning Unit, (1970) *Primary Medical Care*. London: British Medical Association.
British Medical Journal editorial (1959) 'Staphlococcal infections in hospital', *British Medical Journal*, i, 218–9.
British Medical Journal editorial (1960) 'Prevention of harm to patients', *British Medical Journal*, i, 1797–8.
British Medical Journal editorial (1963) 'Distress in dying', *British Medical Journal*, 2: 400–1.
British Medical Journal editorial (1970) 'The name of the game', *British Medical Journal*, ii, 5–6.
British Medical Journal editorial (1968) 'Declaration of Sydney', *British Medical Journal*, 3: 449.
Britten, R.H. (1931) 'The physical examination as an instrument of research', *Public Health Reports*, 46: 1671–76.

Bromberg, W. and Schilder, P. (1936) 'Attitudes of psycho-neurotics to death', *Psychoanalytic Reviews,* 23: 1–25.
Brown, M.A. (1929) *Teaching Health in Fargo*. New York: Commonwealth Fund.
Browne, K. and Freeling, P. (1976) *The Doctor–Patient Relationship*. Edinburgh: Churchill Livingstone.
Brunton, L. (1915) *Collected Papers on Physical Military Training*. London.
Byrne, P.S. and Long, B.E.L (1976) *Doctors talking to patients: a Study of the Verbal Behaviour of General Practitioners Consulting in Their Surgeries*. London: HMSO.
Cabot, R.C. (1905) *Physical Diagnosis*. London: Bailliére Tindall.
Cabot, R.C. and Adams, F.D. (1938) *Physical Diagnosis*. London: Bailliere Tindall, 12th edn.
Campbell, R.B. (1940) 'The psychological aspect of physical education. *Edinburgh Medical Journal,* 47: 351–6.
Caprio, F.S. (1946) 'Ethnologic attitudes to death', *Journal of Clinical Psychopathology,* 7: 737–52.
Carr, E.H. (1961) *What Is History?* London: Macmillan.
Carr-Saunders, A.M and Wilson, P.A. (1933) *The Professions*. London: Cass.
Carson, R. (1962) *Silent spring*. Boston: Houghton Mifflin.
Cartwright, A. (1967) *Patients and Their Doctors*. London: Routledge.
Cartwright, A. and Anderson R. (1981) *General Practice Revisited*. London: Tavistock.
Cohen, H. (1949) Foreword to Seward, C., *Bedside diagnosis*. London: Churchill Livingstone.
Chambers, R. and Maxwell, R. (1996) 'Helping sick doctors', *British Medical Journal*, 312, 722.
Chisholm, C. (1925) 'The physical training of adolescent girls', *Journal of State Medicine* 33: 37–42.
Clarke, F. (1984) *Hospital At home: an Alternative to General Hospital admission*. London: Macmillan–now Palgrave.
Clausen, J.A. Seiden Fold, M.A. and Deasy L.C. (1954) Parent attitudes towards participation of their children in polio vaccine trials. *American Journal of Public Health,* 44: 1526–36.
Cohen, J. (1968) 'Weighted kappa: nominal scale agreement with provision for scaled disagreement and partial credit', *Psychological Bulletin,* 70: 213-20.
Cole, L. (1946) 'Prognosis and the patient', *Lancet* 1: 1–3.
College of General Practitioners, (1962) *Morbidity Statistics from General Practice*. London: HMSO.
Collins, S.D. (1933) 'Causes of illness in 9000 families based on nationwide periodic canvasses, 1928–31, *Public Health Reports,* 48: 283–308.
Collins, S.D. (1934) 'Frequency of health examinations in 9000 families based on nationwide periodic canvasses, 1928–1931', *Public Health Reports,* 49: 321–46.
Cornell, W.S. (1912) *Health and Medical Inspection of School Children*. Philalelphia: Davis.
Cove Smith, R. (1940) 'Factors in national fitness', *Journal of the Royal Institute of Public Health and Hygiene,* 3: 157–61.

Crombie, D.L. (1968) 'Preventive medicine and presymptomatic diagnosis', *Journal of the Royal College of General Practitioners,* 15: 344–51.
Currie, J.R. (1938) *Hygiene*. London: Livingstone.
Dalkey, N.C and Helmar, X. (1963) 'An experimental application of the Delphi method to the use of experts', *Management science,* 9: 458–67.
Davidson, G.W. (1978) *The Hospice: Development and Administration*. Washington, DC: Hemisphere.
Davis, A. and Horobin, G. (eds) (1977) *Medical Encounters*. New York: Croom Helm.
Dawkins, R. (1976) *The Selfish Gene*. Oxford: Oxford University Press.
Dawson Report (1920) *Interim Report on the Future Provision of Medical and Allied Services*. London: HMSO.
De Chaumont, FSBF (undated) *The Habitation in Relation to Health*. London: Society for Promoting Christian Knowledge.
De Chaumont, FSBF (ed.) (1887) *Parkes' Manual of Practical Hygiene*. Philadelphia: Blakiston 7th edn.
Dearing, W.P. (1953) 'New orientation in the teaching of preventive medicine', *Public Health Reports,* 68: 1147–55.
Delbecq, A. and Van de Ven, A. (1971) 'A group process model for problem identification and programme planning', *Journal of Applied Behavioural Science,* 7: 467–92.
Department of Health (1991) *Health and Personal Social Services Statistics*. London: HMSO.
Department of Health (1992) *Health and Personal Social Services Statistics*. London: HMSO.
Derryberry, M. (1945) 'Health education in the public health program', *Public Health Reports* 60: 1394–402.
Derryberry, M. (1949) 'Health is everybody's business', *Public Health Reports* 64: 1293–98.
Douglas, M. (1966) *Purity and Danger*. London: Routledge Kegan Paul.
Downing, A.B. (1969) *Euthanasia and the Right to Death*. New York: Humanities Press.
Draper, P. (ed.) (1991) *Health through public policy*. London: Green Print.
Dukes, CE. (1947) 'The management of permanent colostomy', *Lancet,* 2: 12–14.
Durkheim, E (1912 tran. 1915) *The Elementary Forms of the Religious Life*. London: Allen and Unwin.
Durkheim, E. (1893 tran. 1933) *The Division of Labour in Society*. New York: Free Press.
Egbert, S. (1903) *A Manual of Hygiene and Sanitation*. Philadelphia: Lea Bros, 3rd edn.
Eisenberg, L. (1977) 'Disease and illness: distinctions between professional and popular ideas about illness', *Culture, Medicine and Psychiatry,*1: 9–23.
Elmer, W.P. and Rose, W.D. (1940) *Physical Diagnosis*, 8th edn. rev. by Walker H. St Louis: Mosby.
Emerson, C.P. (1928) *Physical Diagnosis*. Philadelphia: Lippincott.

Emery, IL. (1962) 'Certification of death by the pathologist', *Proceedings of the Royal Society of Medicine,* 55: 738–40.
Enelow, A.J. and Swisher, S.N. (1972) *Interviewing and Patient Care.* Oxford: Oxford University Press.
Evans, A.D. and Howard, L.G.R. (1930) *The Romance of the British Voluntary Hospital Movement.* London: Hutchinson.
Faber, K. (1923) *Nosology in Modern Internal Medicine.* New York: PB Hoeber.
Farr, W. (1839) *Registrar-General's Annual Report for l837–38.* London: HMSO.
Feifel, H. (1955) 'Attitudes of mentally ill patients towards death', *Journal of Nervous and Mental Disease,* 122: 375–380.
Feifel, H. (1959) *The Meaning of Death.* New York: McGraw-Hill.
Fink, A. Kosecoff, J. Chassin, M. and Brook, R.H. (1984) 'Consensus methods: characteristics and guidelines for use', *American Journal of Public Health,* 74: 979–83.
Fitzpatrick, R., Hintsu, J., Newman, S. Scambler, G. and Thompson, J. (1984) *The Experience of Illness.* London: Tavistock.
Fletcher, C.M. (1973) *Communication in medicine.* London: Nuffield.
Florey, C.D., Seuter, M.G. and Acheson, R.M. (1969) A study of the validity of the diagnosis of stroke in mortality data', *American Journal of Epidemiology,* 89: 15–24.
Foucault, M. (1970) *The Order of Things.* London: Tavistock.
Foucault, M. (1973) *The Birth of the Clinic: An archeology of Medical Perception.* London: Tavistock.
Foucault, M. (1977) 'What is an author?', Reprinted in: Bouchard, D. and Simon, S. (eds) *Language, Counter-memory and Practice: Selected Essays and Interviews.* Ithaca, New York: Cornell University press.
Fox, C.B. (1878) *Sanitary Examination of Water, Air and Food: a Handbook for the MOH.* London: Churchill.
Frazer, A.D. (1932) 'The problem of the defaulter', *British Journal of Venereal Diseases* 8: 56–8.
Freidson, E. (1961) *Patients' Views of Medical Practice.* New York: Russell Sage.
Freidson, E. (1970) *Profession of Medicine.* New York: Dodd Mead.
Frisby, B.R. (1957) 'Cleansing of hospital blankets', *British Medical Journal,* ii, 506–8.
Gibson, A.G. and Collier, W.T. (1927) *The Methods of Clinical Diagnosis.* London: Edward Amold.
Gilchrist, A.R. (1960) 'Problems in the management of acute myocardial infarction', *British Medical Journal,* i, 215–19.
Gillie Report (1963) *The Field of Work of' the Family Doctor.* London: HMSO.
Glaser, B.G. and Strauss, A.L. (1965) *Awareness of Dying.* Chicago: Aldine.
Glaser, B.G. and Strauss, A.L. (1968) *Time for Dying.* Chicago: Aldine.
Goffman, E. (1961) *Stigma.* London: Penguin.
Goode, W.J. (1960) 'Encroachment, charlatanism and the emerging profession: psychiatry, sociology and medicine', *American Sociological Review,* 25: 902–14.
Greene, C. (857) *The Burial Acts from 1852 to 1857.* London: Mitchener.

Gulick, L.H. and Ayers, L.P. (1908) *Medical Inspection of Schools*. New York: Russell Sage.
Guy, WA. (1870) *Public Health: a Popular Introduction to Sanitary Science*. London: Renshaw.
Hamer, W.H. (1902) *Manual of Hygiene*. London: Churchill.
Hamlin, H. (1964) Life or death by electroencephalogram. *Journal of the American Medical Association,* 190: 112–14.
Handfield-Jones, R.P.C. (1958) 'The organisation and administration of a general practice', *Journal of the College of General Practitioners,* 1: 205–24.
Handfield-Jones, R.P.C. (1959) 'One year's work in a country practice', *Journal of the College of General Practitioners,* 2: 323–37.
Hanschell, H.M. (1935) 'The defaulting seaman', *British Journal of Venereal Diseases,* 11: 28–30.
Harries, E.H.R. and Mitman, M. (1940) *Clinical Practice in Infectious Diseases*. Edinburgh: Livingstone.
Harris, D.M. and Guten, S. (1979) 'Health protective behaviour an exploratory study', *Journal of Health and Social Behaviour,* 20: 17–29.
Harte, J.D. (1973) 'The long approach to general assessment', *Journal of the Royal College of General Practitioners,* 23: 811–14.
Haviland, A. (1855) *Climate, Weather and Disease*. London: Churchill.
Heasman, M.A. (1962) 'Accuracy of death certification', *Proceedings of the Royal Society of Medicine,* 55: 733–6.
Hicks, D. (1976) *Primary Health Care*. London: HMSO.
Hill, H.W. (1916) *The New Public Health*. New York: Macmillan.
Hilton, J. (1892) *On the Influence of Mechanical and Physiological Rest in the Treatment of Accidents and Surgical Diseases, and the Diagnostic Value of Pain*. 5th edn. London: Bell and Daldy.
Hinton, J. (1963) 'The physical and mental distress of the dying', *Quarterly Journal of Medicine,* 32: 1–21.
Hochbaum, G.M. (1956) 'Why people seek diagnostic X-rays', *Public Health Reports* 71: 377–80.
Hodgkin, K. (1966) *Towards Earlier Diagnosis*. Edinburgh: Livingstone.
Holt, H.M. (1928) 'Physical efficiency: its development in relation to industrial employment', *Journal of State Medicine* 38: 396–405.
Horder, J.P. and Swift, G. (1979) The history of vocational training for general practitioners', *Journal of the Royal College of General Practitioners*. 29: 24
Horder, J. and Horder, E. (1954) 'Illness in general practice', *Practitioner,* 173: 177–94.
Horder, T. and Gow, A.E. (1928) *The Essentials of Medical Diagnosis*. London: Cassell.
Hughes, D.M. (1958) '25 years in country practice. *Journal of the College of General Practitioners,* 1: 5–22.
Hull, F.M. (1972) 'Diagnostic pathways in general practice', *Journal of the Royal College of General Practitioners,* 22: 241–58.
Hussey, M.M. (1928) 'Character education in athletics', *American Physical Education Review,* 3: 578–81.

Hutchison, R. and Hunter, D. (1949) *Clinical methods*. London: Cassell, 11th edn.

Hutchison. R. and Rainy, H. (1935) *Clinical Methods*. London: Cassel, 10th edn.

Illich, I. (1974) *Medical Nemesis*. London: Caldar-Boyars.

Illingworth, R.S. (1953) *The normal child*. London: Churchill.

Interdepartmental Committee Report (1904) *Report of the Interdepartmental Committee on Physical Deterioration*. London: HMSO.

Interdepartmental Committee on Medical Schools (1944) *Goodenough Report*. London: HMSO

Ireland, M.W. (1920) 'Physical and hygienic benefits of military training as demonstrated by the war', *Journal of the American Medical Association*, 74: 499–501.

Isaacs, B., Gunn, J., McKechan, A., McMillan, I. and Neville, Y. (1971) 'The concept of pre-death', *Lancet*, 1: 1115–18.

Janis, I.L. (1958) *Psychological Stress*. New York: Wiley.

Jeans, W.D. (1965) 'Work study in general practice', *Journal of the College of General Practitioners*, 9: 270–79.

Joint Working Party (1967) *Organisation of Medical Work in Hospitals*. London: HMSO.

Journal of the Royal College of General Practitioners editorial. (1969) 'The medical adviser', *Journal of the Royal College of General Practitioners*, 17: 67–8.

Journal of the Royal College of General Practitioners editorial (1973) 'Continuity of care', *Journal of the Royal College of General Practitioners*, 23: 749–50.

Journal of the Royal College of General Practitioners editorial (1984) 'Record requirements', *Journal of the Royal College of General Practitioners*, 34: 68–9.

Journal of the Royal College of General Practitioners editorial (1978) 'Medical records in general practice', *Journal of the Royal College of General Practitioners*, 28: 521–2.

Kagawa-Singer, M. (1993) 'Redefining health: living with cancer', *Social Science and Medicine*, 37: 295–304.

Keith, R.D. (1918) *Clinical Case-taking: an Introduction to Elementary Clinical Medicine*. London: HK Lewis.

King's Fund (1958) *Noise Control in Hospitals*. London: King's Fund.

Kleinman, A., Eisenberg, L. and Good, B. (1978) 'Culture, illness and cure'. *Annals of Internal Medicine*, 88: 251–9.

Kogan, M. Cang, S, Dixon, M. and Tolliday, H. (1971) *Working Relationships within the British Hospital Service*. London: Bookstall.

Koos, E.L. (1954) *The Health of Regionsville: What the People Thought and Did About It*. New York: Hafner.

Kubler-Ross, E. (1969) *On Death and Dying*. New York: Macmillan.

Kubler-Ross, E. (1981) *Living with Death and Dying*. New York: Macmillan.

Kuhn, TS. (1972) *The Structure of Scientific Revolutions*. Chicago: Chicago University Press.

Lancet editorial. (1963) 'Distress of the dying', *Lancet*, 1: 927–8.

Lancet editorial. (1970) 'Why collect statistics?', *Lancet*, 2: 1072–3.

Last, J.M. (1963) 'The iceberg: completing the clinical picture in general practice', *Lancet*, 2: 28–31.
Latour, B. and Woolgar, S. (1979) *Laboratory Life*. Sage.
Lawrence, C. and Shapin, S. (eds) (1998) *Science Incarnate*. Chicago: University of Chicago Press.
Leighton, D.C., Harding, J.S., Macklin, D.B., Macmillan, A.M. and Leighron, A.H. (1963) *The Character of Danger*. New York: Free Press.
Ley, P. (1976) 'Towards better doctor–patient communication'. In: Bennett, SE (ed.), *Communication Between Doctors and Patients*. London: Nuffield.
Lloyd, W.M. (1936) 'Health and physical education', *Medical Officer*, 55: 207–8.
Lorenz, K. (1966) *On aggression*. London: Methuen.
Lowbury, E.J.L. and Lilly, H.A. (1958) 'Contamination of operating-theatre air with Clostridium tetani', *British Medical Journal*, ii, 1335–6.
MacFarlane, W.V. and Johns, H.M. (1947) 'The problem of default in a venereal disease clinic: a medico-social analysis of 381 women patients', *British Journal of Venereal Diseases*, 23: 171–9.
Mackenzie, W.L. (1906) *The Health of the School Child*. London: Methuen.
Martin, C.J. and McQueen, D.V. (1989) *Readings for a New Public Health*. Edinburgh: Edinburgh University Press.
Mauss, M. (1925) Trans. by Cunnison, I (1954), *The Gift*. London.
McCleary, GF. (1936) *The Maternity and Child Welfare Movement*. London: King.
McLeod, J. (ed.) (1973) *Clinical Examination*, 3rd edn. London: Churchill Livingstone.
McLeod, J. (ed.) (1976) *Clinical Examination*. London: Churchill Livingston, 4th edn.
Noble Chamberlain, E. (1938) *Symptoms and Signs in Clinical Medicine*. Bristol: Wright, 3rd edn.
McWhinney, I.R. (1964) *The Early Signs of Illness*. London: Pitman.
Meares, A. (1960) 'Communication and the patient', *Lancet*, 1: 663–7.
Mechanic D. and Volkart, E.H. (1960) 'Illness behaviour and medical diagnoses', *Journal of Health and Social Behaviour*, 1: 86–94.
Medical Services Study Group (1978) 'Death certification and epidemiological research', *British Medical Journal*, 2: 1063–5.
Melzack, R. and Wall, P.D. (1965) 'Pain mechanisms: a new theory', *Science*, 150, 971–9.
Merton, R.K., Reader, G.G. and Kendall, P.L. (1957) *The Student Physician*. Cambridge: Harvard University Press.
Miers, H.A. and Cross Key, R. (1893) *The Soil in Relation to Health*. London: Macmillan.
Milligan, W. (1918) 'The value and importance of physical exercise from a national standpoint', *Journal of State Medicine*, 26: 33–43.
Ministry of Health (1959) *The Welfare of Children in Hospitals*. (Platt Committee). London: HMSO.
Ministry of Health (1961a) *Pattern of the In-patients Day*. London: HMSO.
Ministry of Health (1961b) *Human Relations in Hospital*, London: HMSO.
Ministry of Health (1967) *Buildings for General Medical Practice*. London: HMSO.
Morison, R.S. (1971) 'Death: process or event?', *Science*, 173: 694–8.

Morris, D. (1967) *The Naked Ape*. London: Cape.
Morris, D. (1969) *The Human Zoo*. London: Cape.
Nabarro, D. (1935) 'The defaulting child', *British Journal of Venereal Diseases*, 11: 91–7.
Neighbour, R. (1987) *The Inner Consultation: How to Develop an Effective and Intuitive Consulting Style*. Lancaster: MTP Press.
Newman, G. (1906) *Infant Mortality: a Social problem*. London: Methuen.
Newman, G. (1919) *An Outline of the Practice of Preventive Medicine*. London: HMSO.
Newsholme, A. (1892) *Hygiene: a Manual of Personal and Public Health*. London: Gill.
Newton, RC (1907) 'What should be the attitude of the profession towards the hygiene of school life?', *Journal of the American Medical Association*, 49: 663–5
Nichol, H. (1935a) 'The defaulting prostitute', *British Journal of Venereal Diseases*, 11: 31–5
Nichol, H. (1935b) 'The defaulting travelling man', *British Journal of Venereal Diseases*, 11: 98–104.
Nightingale, F. (1859) *Notes on Nursing*. London: Harrison.
Noble Chamberlain, E. (1938) *Symptoms and Signs in Clinical Medicine*. Bristol: Wright, 2th edn.
Noble Chamberlain, E. (1952) *Symptoms and Signs in Clinical Medicine*. Bristol: Wright, 5th edn.
Noble Chamberlain, E. (1957) *Symptoms and Signs in Clinical Medicine*. Bristol: Wright, 6th edn.
Noble Chamberlain, E. (1961) *Symptoms and Signs in Clinical Medicine*. Bristol: Wright, 7th edn.
Noble Chamberlain, E. and Ogilvie, C.M. (1967) *Symptoms and Signs in Clinical Medicine*. Bristol: Wright, 8th edn.
Ogilvie, H. (1957) 'Journey's end', *Practitioner*, 179: 584–91.
Oleson, S.D. (1939) 'What people ask about health', *Public Health Reports*, 54: 765–90.
Parkes, E.A. (1873) *A Manual of Practical Hygiene*. London: Churchill, 4th edn.
Parkin, J. (1859) *The Causation and Prevention of Disease*. London: Churchill.
Parkinson, J. (1951) 'The patient and the physician', *Lancet*, 2: 457–9.
Parsons, T. (1951) *The Social System*. New York: Free Press.
Pearse, I.H. and Crocker, L.H. (1943) *The Peckham Experiment*. London: Allen and Unwin.
Pendleton, D., Schofield, T., Tate, P. and Havelock, P. (1984) *The Consultation: an Approach to Learning and Teaching*. Oxford: Oxford University Press.
Phillips, H.J. (1934) 'A note on flat foot: being a plea for the introduction of remedial exercises in elementary schools', *Medical Officer*, 52: 55–6.
Pickering, G. (1962) 'Logic and hypertension', *Lancet*, 2: 149–50.
Pinker, R. (1968) *English Hospital Statistics 1861–1938*. London: Heinemann.
Pinsent, R.J.F.H. (1969) 'Continuing care in general practice', *Journal of the Royal College of General Practitioners*, 17: 223–6.

Platt, R. (1963) 'Reflections on ageing and death', *Lancet* 1: 1–8.
Poore, G.V. (1902) *The Earth in Relation to the Preservation and Destruction of Contagia*. London: Longman Green.
Porter, C. (1906) *School Hygiene and the Laws of Health*. London: Longman Green.
Practitioner (1953) 170: 567.
Raynes, N.V. and Cairns, V. (1980) 'Factors contributing to the length of general practice consultations', *Journal of the Royal College of General Practitioners*, 30: 496–8.
Registrar-General, (1839) *Annual Report for 1837–38*. London: HMSO.
Registrar-General, (1857) *Annual Report for 1855*. London: HMSO.
Registrar-General, (1869) *Annual Report for 1867*. London: HMSO.
Registrar-General, (1870) *Annual Report for 1868*. London: HMSO.
Registrar-General, (1881) *Annual Report for 1879*. London: HMSO.
Registrar-General, (1887) *Annual Report for 1885*. London: HMSO.
Registrar-General, (1905) *Annual Report for 1903*. London: HMSO.
Registrar-General, (1907) *Annual Report for 1905*. London: HMSO.
Registrar-General, (1909) *Annual Report for 1908*. London: HMSO.
Registrar-General, (1918) *Annual Report for 1916*. London: HMSO.
Registrar-General, (1938) *Statistical Review for 1937*. London: HMSO.
Registrar-General, (1947) *Statistical Review for 1938–9*. London: HMSO.
Registrar-General, (1949) *Statistical Review for 1948*. London: HMSO.
Registrar-General, (1954) *Statistical Review for 1952*. London: HMSO.
Reports from General Practice. (1965) 'Present state and future needs', *Journal of the College of General Practitioners*, Suppl.
Report of a symposium on early diagnosis. (1967) *Journal of the College of General Practitioners*, Suppl.
Report of a Working Party (1981) *Health and Prevention in Primary Care*. London: Royal College of General Practitioners.
Research Committee (1958) 'The continuing observation and recording of morbidity', *Journal of the College of General Practitioners*, 1: 107–28.
Richard, J. (1962) 'The Stranraer health centre', *Journal of the College of General Practitioners*, 5: 256–64.
Riska, E. (2000) 'The rise and fall of Type A man', *Social Science and Medicine*, 51: 1665–74.
Rosenstock, I.M., Derryberry, M. and Carriger, B.K. (1959) 'Why people fail to seek poliomyelitis vaccination', *Public Health Reports*, 74: 98–103.
Rosenstock, I.M. (1965) 'Why people use health services', *Millbank Memorial Fund Quarterly*, 44: 94–127.
Roth, J. (1963) *Timetables*. Indianapolis: Bobbs-Merrill.
Rotter, J.B. (1966) 'Generalized expectancies for the internal versus external control of reinforcement', *Psychological Monographs*, 90: 1–28.
Royal College of General Practitioners (1972) *The Future General Practitioner*. London: Royal College of General Practitioners.
Ruble, V.W. (1928) 'A psychological study of athletes', *American Physical Education Review*, 33: 216–9.
Sand, R. (1952) *The Advance to Social Medicine*. New York: Staple Press.

Saunders, C. and Baines, M. (1983) *Living with Dying*. Oxford: Oxford University Press.

Schenthal, J.E. (1960) 'Multiphasic screening of the well patient', *Journal of the American Medical Association*, 172: 51–64.

Schneidman, E.S. (1976) *Death: Current perspectives*. Palo Alto: Mayfield.

Schwenk, T.L. and Hughes, C.C. (1983) 'The family as patient in family medicine: rhetoric or reality?', *Social Science and Medicine*, 17: 1–16.

Scott, R. (1965) 'Medicine in society', *Journal of the College of General Practitioners*, 9: 3–16.

Selye, H. (1956) *Stress*. New York: McGraw-Hill.

Simpson, S.L. (1937) *Medical Diagnosis: Some Clinical Aspects*. London: HK Lewis.

Smith, R. (1997) 'All doctors are problem doctors', *British Medical Journal*, 314, 841–2.

Smith, S. (1866) *The Common Nature of Epidemics and Their Relation to Climate and Civilisation*. London: Trubner.

Srole L. Langnet, T.S., Michael, S.T., Opler, M.K. and Rennie, T.A.C. (1962) *Mental Health in the Metropolis*. New York: McGraw-Hill.

Standard, S. and Nathan, H. (1955) *Should the Patient Know the Truth?* New York: Springer.

Standing Medical Advisory Committee of the Central Health Services Council (1959) *Staphlococcal infections in hospitals*. London: HMSO.

Standing Medical Advisory Committee (1971) *The Organisation of General Practice*. London: HMSO.

Stern, N.S. (1933) *Clinical Diagnosis: Physical and Differential*. New York: Macmillan.

Stevens, W. (1910) *Medical diagnosis*. London: H.K Lewis.

Stimson, G. and Webb, B. (1975) *Going to See the Doctor: the Consultation Process in General Practice*. London: Routledge and Kegan Paul.

Stott, P. (1983) *Milestones: the Diary of a Trainee GP*. London: Pan.

Sub-Committee of the Central Health Services Council. (1963) *Communication between Doctors, Nurses and Patients: an Aspect of Human Relations in the Hospital Service*. London: HMSO.

Sutherland, R. (1950) 'Some individual and social factors in venereal disease', *British Journal of Venereal Diseases*, 25: 1–12.

Szasz, T.S. (1962) *The Myth of Mental Illness*. St Albans: Palladin.

Taylor, C.H.S. (1934) 'Physical exercises as a means of preserving health', *The Practitioner*, 132: 280–89.

Taylor, S. (1954) *Good General Practice*. London: Oxford University Press.

The Organisation of Group Practice. (1971) London: HMSO.

Thomas, C.G.A., Liddell, J. and Carmichael, D.S. (1958) *British Medical Journal*, ii, 1336–8.

Tinbergen, N. (1953) *Social Behaviour in Animals with Special Reference to Vertebrates*. New York: Wiley.

Townsend, E. (1962) 'Future trends in general practice', *Journal of the College of General Practitioners*, 5: 501–24.

Tredgold, R.F. (1962) 'The integration of psychiatric teaching in the curriculum', *Lancet*, 1: 1344–7.

Treloar, A.E. (1956) 'The enigma of cause of death', *Journal of the American Medical Association*, 162: 1377–9.
Tuckett, D. Boulton, M. Olsen, C. and Williams, A. (1985) *Meetings Between Experts*. London: Tavistock.
Wadsworth, M. and Robinson, D. (eds) (1976) *Studies in Everyday Medical Life*. London: Martin Robertson.
Walker, G.A. (1839) *Gatherings from Graveyards*. London: Longman.
Watson, C. and Clarke, M. (1972) 'Attachment schemes and development of the health care team', *Update*, 5: 489–94.
Westcott, P. (1977) 'The length of consultations in general practice', *Journal of the Royal College of General Practitioners*, 27: 552–5.
Whitaker, A.J. (1965) 'A study of purpose built group practice premises', *Journal of the College of General Practitioners*, 10: 265–71.
Wiesman, A.D. (1972) *On Death and Denying*. New York: Behavioural Publications.
Williams, R.E.O. and Shooter, R.A. (1963) *Infection in Hospitals: Epidemiology and Control. UNESCO/WHO Symposium*. Oxford: Blackwell.
Williams, R.E.O. Blowers, R. Garrod, L.P. and Shorter, R.A. (1960) *Hospital Infection: Causes and Prevention*. London: Lloyd-Luke.
Williams, R.E.O. Jevons, M.P. Shooter, R.A., Hunter, C.J.W., Gi Ming, J.A., Griffiths, J.D. and Tagler, G.W. (1959) 'Nasal staphlococcus and sepsis in hospital patients', *British Medical Journal*, ii, 658–62.
Wilson, G. (1892) *A Handbook of Hygiene and Sanitary Science*. Philadelphia: Blakiston.
Wilson, C. (1956) *The outsider*. London: Gollanz.
Wishik, GM. (1958) 'Attitudes and reactions of the public to health programs', *American Journal of Public Health*, 48: 139–41.
Wittkower, E.D. (1948) 'Psychological aspects of venereal disease', *British Journal of Venereal Diseases*, 24: 59–67.
Wood, L.A.C. (1962) 'A time and motion study', *Journal of the College of General Practitioners*, 5: 379–81.
Working Group on Terminal Care. (1980) 'National terminal care policy', *Journal of the Royal College of General Practitioners*, 30: 466–71.
Zola, I.K. (1973) 'Pathways to the doctor: from person to patient', *Social Science and Medicine*, 7: 677–89.
Zola, I.K. (1972) 'Medicine as an institution of social control', *Sociological Review*, 20: 487–504.

Index